REVISE EDEXCEL AS
Mathematics

REVISION GUIDE

Series Consultant: Harry Smith

Author: Harry Smith

Notes from the publisher

While the publishers have made every attempt to ensure that advice on the qualification and its assessment is accurate, the official specification and associated assessment guidance materials are the only authoritative source of information and should always be referred to for definitive guidance.

Pearson examiners have not contributed to any sections in this resource relevant to examination papers for which they have responsibility.

Also available to support your revision:

Revise A Level Revision Planner   9781292191546

The **Revise A Level Revision Planner** helps you to plan and organise your time, step-by-step, throughout your A level revision. Use this book and wall chart to mastermind your revision.

For the full range of Pearson revision titles across KS2, KS3, GCSE, Functional Skills, AS/A Level and BTEC visit:
www.pearsonschools.co.uk/revise

Contents

MECHANICS

. . . .

A small bit of small print: Edexcel publishes Sample Assessment Material and the Specification on its website. This is the official content and this book should be used in conjunction with it. The questions have been written to help you practise every topic in the book.

Index laws

You need to be able to work with algebraic expressions confidently for all of your AS maths topics. Make sure you know how to use these six index laws.

1 $\quad a^m \times a^n = a^{m+n} \qquad x^5 \times x^{-2} = x^3$

2 $\quad \dfrac{a^m}{a^n} = a^{m-n} \qquad \dfrac{x^8}{x^6} = x^2$

3 $\quad (a^m)^n = a^{mn} \qquad (x^4)^3 = x^{12}$

4 $\quad a^{-n} = \dfrac{1}{a^n} \qquad 5^{-2} = \dfrac{1}{5^2} = \dfrac{1}{25}$

5 $\quad a^{\frac{1}{n}} = \sqrt[n]{a} \qquad 49^{\frac{1}{2}} = 7$

$\qquad\qquad\qquad\qquad 27^{\frac{1}{3}} = 3$

$\quad 2 \times 2 \times 2 \times 2 = 16 \qquad 16^{\frac{1}{4}} = 2$
$\quad$ so $\sqrt[4]{16} = 2$

6 $\quad a^{\frac{m}{n}} = (\sqrt[n]{a})^m \qquad 27^{-\frac{2}{3}} = \left(27^{\frac{1}{3}}\right)^{-2}$

$\qquad\qquad\qquad\qquad = \left(\sqrt[3]{27}\right)^{-2}$

Do these calculations one step at a time.

$\qquad\qquad\qquad\qquad = 3^{-2} = \dfrac{1}{3^2} = \dfrac{1}{9}$

Worked example

Given that $\dfrac{10x - 5x^{\frac{7}{2}}}{\sqrt{x}}$ can be written in the form $10x^p - 5x^q$, write down the value of p and the value of q.   **(2 marks)**

$\dfrac{10x}{x^{\frac{1}{2}}} - \dfrac{5x^{\frac{7}{2}}}{x^{\frac{1}{2}}} = 10x^{1-\frac{1}{2}} - 5x^{\frac{7}{2}-\frac{1}{2}}$

$\qquad\qquad\qquad = 10x^{\frac{1}{2}} - 5x^3$

$p = \dfrac{1}{2}$ and $q = 3$

The fraction acts like a bracket, so you need to divide **both** terms in the numerator by $\sqrt{x}$. Write down the values of p and q when you've finished.

Golden rule

Convert all roots into fractional powers before applying the other index laws.

$\sqrt{x} = x^{\frac{1}{2}} \qquad \sqrt[3]{x} = x^{\frac{1}{3}}$

You will be allowed to use your calculator in both of your AS exams, but be careful. If you enter $125^{-\frac{1}{3}}$ into your calculator it will give you the answer $\frac{1}{5}$ or 0.2. You need to be able to write it in the form given in the question.

Check that your answer is in the correct form. n is an integer so it is a positive or negative whole number, or 0.

Worked example

Express $125^{-\frac{1}{3}}$ in the form 5^n where n is an integer.   **(1 mark)**

$125^{-\frac{1}{3}} = (5^3)^{-\frac{1}{3}}$

$\qquad\quad = 5^{3 \times \left(-\frac{1}{3}\right)}$

$\qquad\quad = 5^{-1}$

So $n = -1$

Now try this

1 Simplify $x\left(4x^{-\frac{1}{2}}\right)^3$ **(2 marks)**

2 Simplify $\left(9y^{10}\right)^{\frac{3}{2}}$ **(2 marks)**

There are full worked solutions to all the Now try this questions on page 96.

3. Write $\dfrac{5 + 2\sqrt{x}}{x^2}$ in the form $5x^p + 2x^q$, where p and q are constants.   **(2 marks)**

Remember that a constant doesn't have to be an integer.

Expanding and factorising

In your AS exam, you might have to **multiply out** (or **expand**) a product of **three** brackets, or **factorise** a **cubic** expression.

Expanding brackets

To expand the product of two factors, you have to multiply **every term** in the first factor by **every term** in the second factor.

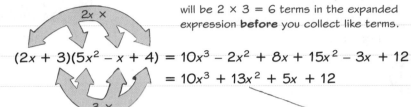

There are 2 terms in the first factor and 3 terms in the second factor, so there will be $2 \times 3 = 6$ terms in the expanded expression **before** you collect like terms.

$$(2x + 3)(5x^2 - x + 4) = 10x^3 - 2x^2 + 8x + 15x^2 - 3x + 12$$
$$= 10x^3 + 13x^2 + 5x + 12$$

Simplify your expression by collecting like terms: $-2x^2 + 15x^2 = 13x^2$

Remember that $\sqrt{x} \times \sqrt{x} = x$.
$(2\sqrt{x})^2 = 2\sqrt{x} \times 2\sqrt{x} = 4\sqrt{x}\sqrt{x} = 4x$

Worked example

Show that $(3 + 2\sqrt{x})^2$ can be written as $9 + k\sqrt{x} + 4x$, where k is a constant to be found. **(2 marks)**

$(3 + 2\sqrt{x})(3 + 2\sqrt{x}) = 9 + 6\sqrt{x} + 6\sqrt{x} + 4x$
$\qquad\qquad\qquad\qquad = 9 + 12\sqrt{x} + 4x$

$k = 12$

Problem solved!

If you have to find a constant, it's a good idea to write down the value of the constant when you have finished your working.

You will need to use problem-solving skills throughout your exam – **be prepared!**

Factorising

Factorising is the opposite of expanding brackets.

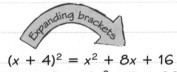

$(x + 4)^2 = x^2 + 8x + 16$
$(2x + 3)(x - 10) = 2x^2 - 7x - 30$

Worked example

Factorise completely $x^3 + x^2 - 6x$ **(3 marks)**

$x^3 + x^2 - 6x = x(x^2 + x - 6)$
$\qquad\qquad\qquad = x(x + 3)(x - 2)$

Start by taking the common factor, x, out of every term. You are left with a quadratic expression, which you can factorise into two linear factors.

Now try this

1 Given that $(x + 2)(x + 1)^2 = x^3 + bx^2 + cx + d$, where b, c and d are constants, find the values of b, c and d.   **(3 marks)**

Multiply out $(x + 1)^2$ first to get $(x + 2)(x^2 + 2x + 1)$.

2 Factorise completely $3x^3 - 2x^2 - x$ **(3 marks)**

3 Factorise completely $25x^2 - 16$ **(2 marks)**

This is a difference of two squares: $a^2 - b^2 = (a + b)(a - b)$

2

Surds

You can simplify surds quickly using the fraction and square root functions on your calculator. However, you need to understand how to manipulate surds in order to work with algebraic expressions confidently.

Golden rules

These are the two golden rules for simplifying surds:

1 $\sqrt{ab} = \sqrt{a} \times \sqrt{b}$

$\sqrt{8} = \sqrt{4} \times \sqrt{2} = 2\sqrt{2}$

2 $\sqrt{\dfrac{a}{b}} = \dfrac{\sqrt{a}}{\sqrt{b}}$ $\sqrt{\dfrac{3}{25}} = \dfrac{\sqrt{3}}{\sqrt{25}} = \dfrac{\sqrt{3}}{5}$

Worked example

Write $\sqrt{45}$ in the form $k\sqrt{5}$, where k is an integer. **(2 marks)**

$\sqrt{45} = \sqrt{9 \times 5}$

$\phantom{\sqrt{45}} = \sqrt{9} \times \sqrt{5}$

$\phantom{\sqrt{45}} = 3\sqrt{5}$

$k = 3$

Worked example

Simplify $\sqrt{27} + \sqrt{48}$, giving your answer in the form $a\sqrt{3}$, where a is an integer. **(2 marks)**

$\sqrt{27} + \sqrt{48} = \sqrt{9} \times \sqrt{3} + \sqrt{16} \times \sqrt{3}$

$\phantom{\sqrt{27} + \sqrt{48}} = 3\sqrt{3} + 4\sqrt{3}$

$\phantom{\sqrt{27} + \sqrt{48}} = 7\sqrt{3}$

$\phantom{\sqrt{27} + \sqrt{48}} a = 3$

$\sqrt{a} + \sqrt{b}$ is **not** equal to $\sqrt{a + b}$.

You know you need to write the answer in the form $a\sqrt{3}$ so write each surd in the form $k\sqrt{3}$. You need to take a factor of 3 out of each number:

$27 = 9 \times 3$ so $\sqrt{27} = \sqrt{9 \times 3}$

$48 = 16 \times 3$ so $\sqrt{48} = \sqrt{16 \times 3}$

Rationalising the denominator

You can rationalise the denominator of a fraction by removing any surds in the denominator.

$$\overset{\times (4 + \sqrt{11})}{\overbrace{\dfrac{1}{4 - \sqrt{11}}}} = \dfrac{4 + \sqrt{11}}{(4 - \sqrt{11})(4 + \sqrt{11})} = \dfrac{4 + \sqrt{11}}{5}$$

$\times (4 + \sqrt{11})$

$(4 - \sqrt{11})(4 + \sqrt{11}) = 16 - 4\sqrt{11} + 4\sqrt{11} - 11$

$\phantom{(4 - \sqrt{11})(4 + \sqrt{11})} = 5$

To work out what to multiply the top and bottom by, look at the denominator of the original fraction. Swap a plus for a minus, or swap a minus for a plus.

Worked example

Express $\dfrac{14}{3 + \sqrt{2}}$ in the form $a + b\sqrt{2}$, where a and b are integers. **(2 marks)**

$\dfrac{14}{3 + \sqrt{2}} = \dfrac{14(3 - \sqrt{2})}{(3 + \sqrt{2})(3 - \sqrt{2})}$

$\phantom{\dfrac{14}{3 + \sqrt{2}}} = \dfrac{14(3 - \sqrt{2})}{9 + 3\sqrt{2} - 3\sqrt{2} - 2}$

$\phantom{\dfrac{14}{3 + \sqrt{2}}} = \dfrac{\overset{2}{14}(3 - \sqrt{2})}{\underset{1}{7}}$

$\phantom{\dfrac{14}{3 + \sqrt{2}}} = 6 - 2\sqrt{2}$

$a = 6$ and $b = -2$

If the denominator is in the form $p + \sqrt{q}$ then multiply the numerator and denominator of the fraction by $p - \sqrt{q}$.

Now try this

1   Expand and simplify $(x + \sqrt{3})(x - \sqrt{3})$. **(2 marks)**

2   Write $\sqrt{98}$ in the form $a\sqrt{2}$, where a is an integer. **(1 mark)**

3   Simplify the following, giving your answers in the form $a + b\sqrt{5}$, where a and b are integers.

(a) $\dfrac{8}{3 + \sqrt{5}}$ **(2 marks)**   (b) $\dfrac{4 + \sqrt{5}}{2 - \sqrt{5}}$ **(4 marks)**

Quadratic equations

Quadratic equations can be written in the form $ax^2 + bx + c = 0$, where a, b and c are constants. The solutions of a quadratic equation are sometimes called the **roots** of the equation.

Solution by factorising

You can follow these steps to solve some quadratic equations.

1. **Rearrange** the equation into the form $ax^2 + bx + c = 0$

2. **Factorise** the left-hand side.

3. Set each factor **equal to zero** and solve to find two values of x.

> The first solution is the value of x which makes the $(2x + 3)$ factor equal to 0.

Worked example

Solve $2(x + 1)^2 = 3x + 5$ **(4 marks)**

$$2(x^2 + 2x + 1) = 3x + 5$$
$$2x^2 + 4x + 2 = 3x + 5$$
$$2x^2 + x - 3 = 0$$
$$(2x + 3)(x - 1) = 0$$

$2x + 3 = 0$ or $x - 1 = 0$

$x = -\dfrac{3}{2}$ $x = 1$

Worked example

$x^2 + 6x - 2 = (x + a)^2 + b$, where a and b are constants.

(a) Find the values of a and b.   **(3 marks)**

$$x^2 + 6x - 2 = (x + 3)^2 - 3^2 - 2$$
$$= (x + 3)^2 - 9 - 2$$
$$= (x + 3)^2 - 11$$

$a = 3$ and $b = -11$

(b) Hence, or otherwise, show that the roots of
$$x^2 + 6x - 2 = 0$$
are $c \pm \sqrt{11}$, where c is an integer to be found.   **(2 marks)**

$(x + 3)^2 - 11 = 0$ $(+ 11)$

$(x + 3)^2 = 11$ $(\sqrt{})$

$x + 3 = \pm\sqrt{11}$ (-3)

$x = -3 \pm \sqrt{11}$

$c = -3$

Completing the square

You can write a quadratic expression in the form $(x + p)^2 + q$ using these two identities:

1 $x^2 + 2bx + c \equiv (x + b)^2 - b^2 + c$

2 $x^2 - 2bx + c \equiv (x - b)^2 - b^2 + c$

You can use this method to solve a quadratic equation without using a calculator.

> Write the left-hand side in completed square form, and use inverse operations to solve the equation. Remember that any positive number has **two** square roots: one positive and one negative. You need to use the $\pm$ symbol when you take square roots of both sides of the equation.

Now try this

1 Solve the equation $2(x - 3)^2 + 3x = 14$ **(3 marks)**

2 (a) Show that $x^2 - 10x + 7$ can be written as $(x + p)^2 + q$, where p and q are integers to be found.   **(2 marks)**

> $x^2 - 10x + 7 = (x - 5)^2 - 5^2 + 7$

(b) Hence solve the equation $x^2 - 10x + 7 = 0$, giving your answer in the form $x = a \pm b\sqrt{2}$, where a and b are integers to be found.   **(3 marks)**

> You could also solve this by substituting $a = 1$, $b = -10$ and $c = 7$ into the quadratic formula,
> $$x = \frac{-b \pm \sqrt{b^2 - 4ac}}{2a}, \text{ then}$$
> simplifying.

Functions and roots

You need to be able to use function notation and to be able to solve quadratic equations in a **function of the unknown**. Here is an example:

You don't know how to solve a quartic equation like this, but if you write $u = x^2$ then $u^2 = (x^2)^2 = x^4$.

$$x^4 - 7x^2 + 12 = 0$$
$$u^2 - 7u + 12 = 0$$
$$(u - 3)(u - 4) = 0$$

Substitute $u = x^2$ to form a quadratic equation in u.

$$u = 3 \quad \text{or} \quad u = 4$$

Use the substitution to convert back to x. This equation has **four solutions**.

$$x^2 = 3 \qquad x^2 = 4$$
$$x = \pm\sqrt{3} \qquad x = \pm 2$$

You could also write
$$(x^2)^2 - 7(x^2) + 12 = 0$$
$$(x^2 - 3)(x^2 - 4) = 0$$

Worked example

Solve the equation $3x + \sqrt{x} - 2 = 0$

(4 marks)

Let $u = \sqrt{x}$

$$3x + \sqrt{x} - 2 = 0$$
$$3u^2 + u - 2 = 0$$
$$(3u - 2)(u + 1) = 0$$

$3u - 2 = 0$ or $u + 1 = 0$

$u = \dfrac{2}{3}$ or $\cancel{u = -1}$

Using $u = \sqrt{x}$

$$x = u^2$$
$$x = \frac{4}{9}$$

Problem solved!

$x = (\sqrt{x})^2$ so you could write the equation as:
$$3(\sqrt{x})^2 + \sqrt{x} - 2 = 0$$

This is a **quadratic equation in** $\sqrt{x}$. The safest way to solve equations like this is to use the substitution $y = \sqrt{x}$:

$\sqrt{x}$ is the positive square root of x, so it can only take positive values. You need to ignore the solution $u = -1$.

You will need to use problem-solving skills throughout your exam – **be prepared!**

Domain

Functions will usually be defined for a given domain. This is the set of values that can be used as the input to the function. The domain of this function is all the positive **real numbers** ($\mathbb{R}$).

$$g(x) = 2x^2 - 5x - 3, \ x \in \mathbb{R}, \ x > 0$$

The **roots** of a function, $g(x)$, are the values of x for which $g(x) = 0$. You might need to consider the domain when finding the roots of a function.

Worked example

$g(x) = 2x^2 - 5x - 3, \ x \in \mathbb{R}, \ x > 0$

Show that $g(x)$ has exactly one root and find its value. **(3 marks)**

Set $g(x) = 0$ to find roots:

$$2x^2 - 5x - 3 = 0$$
$$(2x + 1)(x - 3) = 0$$

$x = -\dfrac{1}{2}$ or $x = 3$

The domain is $x > 0$, so $x = -\frac{1}{2}$ is not a root. The only root is $x = 3$.

Now try this

1 Solve
   (a) $x^4 - 3x^2 - 4 = 0$ **(4 marks)**
   (b) $8x^6 + 7x^3 - 1 = 0$ **(4 marks)**
   (c) $x + 10 = 7\sqrt{x}$ **(4 marks)**

2 Solve the equation $4\sqrt{x} + x = 3$, giving your answer in the form $a - b\sqrt{7}$, where a and b are integers to be found. **(5 marks)**

3 $f(x) = x^4 - 4x^2 - 5, \ x \in \mathbb{R}, \ x < 0$
   Show that $f(x)$ has only one root and determine its exact value. **(4 marks)**

Look at the domain of the function carefully. $x^4 - 4x^2 - 5 = 0$ has **two real solutions** but only one of them is a root of $f(x)$.

Sketching quadratics

When you **sketch** a graph you need to show its key features. You don't need to use graph paper for a sketch, but you should still draw your axes and any straight lines using a **ruler**.

Factorised quadratics

On a sketch you usually show the points where the graph crosses the axes.

$y = (x + 2)(x - 5)$

When $x = 0$,
$y = (0 + 2)(0 - 5)$
$= -10$

When $y = 0$,
$(x + 2)(x - 5) = 0$
$x = -2$ or $x = 5$

$y = 4x - x^2$, so the coefficient of x^2 is **negative**.

Negative coefficients

If the coefficient of x^2 is **negative** the graph will be an 'upside down' U-shape. You can check the coefficient of x^2 by multiplying out the brackets.

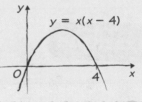

Worked example

Sketch the curve with equation $y = x(4 - x)$
(3 marks)

$y = x(x - 4)$

Worked example

Sketch the curve with equation $y = (x + 2)^2 + 1$
Label the minimum point and any points where the curve crosses the coordinate axes. **(3 marks)**

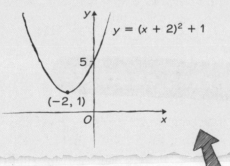

$y = (x + 2)^2 + 1$

5

$(-2, 1)$

To work out the point where the curve crosses the y-axis, substitute $x = 0$ into the equation.

Sketching $y = (x + a)^2 + b$

If a quadratic is written in **completed square** form then you can find the position of its **vertex** easily.

$y = (x + 2)^2 - 5$

-1

$(-2, -5)$

The curve with equation $y = (x + a)^2 + b$ has a vertex at $(-a, b)$.

Now try this

1 On separate diagrams, sketch the graphs of

(a) $y = (x - 2)^2$ **(3 marks)**

(b) $y = (x - 2)^2 + k$, where k is a positive constant.   **(2 marks)**

Show on each sketch the turning point and the coordinates of any points where the graph meets the axes.

2 $x^2 + 6x + 15 = (x + a)^2 + b$

(a) Find the values of a and b.   **(2 marks)**

(b) Sketch the graph of $y = x^2 + 6x + 15$, labelling the minimum point and any points of intersection with the axes.   **(3 marks)**

(c) Use your graph to explain why the equation $x^2 + 6x + 15 = 0$ has no real solutions.   **(1 mark)**

You can also use the discriminant to determine the number of real solutions. There is more about this on the next page.

The discriminant

The discriminant of a quadratic expression $ax^2 + bx + c$ is the value $b^2 - 4ac$. You can use the discriminant to work out whether a quadratic equation has any **real roots** or **real solutions**. There are three possible conditions for the discriminant.

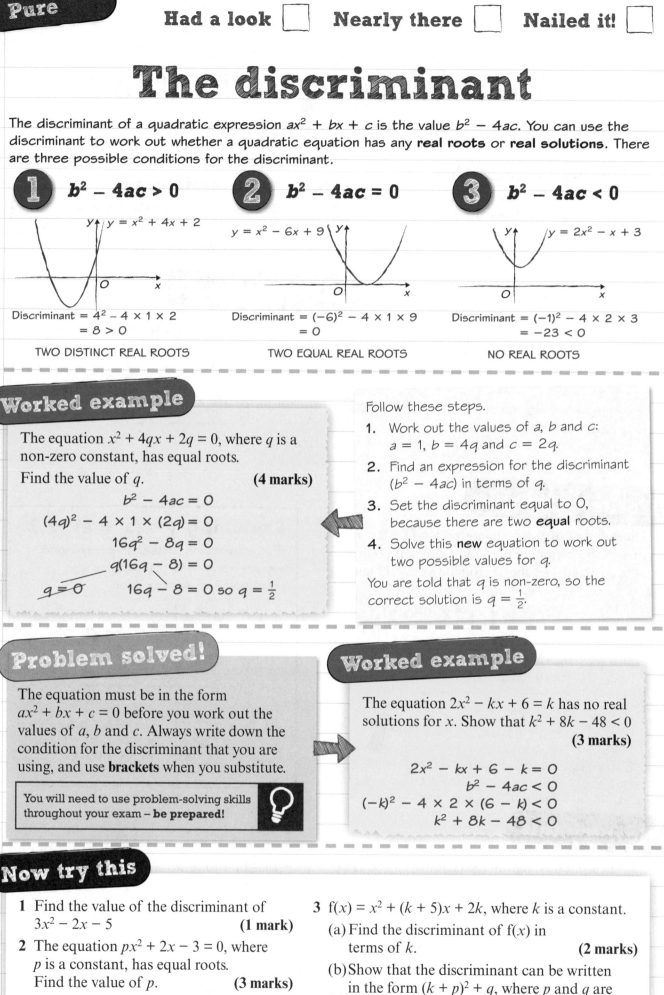

1 $b^2 - 4ac > 0$

$y = x^2 + 4x + 2$

Discriminant $= 4^2 - 4 \times 1 \times 2$
$= 8 > 0$

TWO DISTINCT REAL ROOTS

2 $b^2 - 4ac = 0$

$y = x^2 - 6x + 9$

Discriminant $= (-6)^2 - 4 \times 1 \times 9$
$= 0$

TWO EQUAL REAL ROOTS

3 $b^2 - 4ac < 0$

$y = 2x^2 - x + 3$

Discriminant $= (-1)^2 - 4 \times 2 \times 3$
$= -23 < 0$

NO REAL ROOTS

Worked example

The equation $x^2 + 4qx + 2q = 0$, where q is a non-zero constant, has equal roots.
Find the value of q.   **(4 marks)**

$$b^2 - 4ac = 0$$
$$(4q)^2 - 4 \times 1 \times (2q) = 0$$
$$16q^2 - 8q = 0$$
$$q(16q - 8) = 0$$
$$\cancel{q = 0} \qquad 16q - 8 = 0 \text{ so } q = \tfrac{1}{2}$$

Follow these steps.

1. Work out the values of a, b and c:
   $a = 1$, $b = 4q$ and $c = 2q$.

2. Find an expression for the discriminant ($b^2 - 4ac$) in terms of q.

3. Set the discriminant equal to 0, because there are two **equal** roots.

4. Solve this **new** equation to work out two possible values for q.

You are told that q is non-zero, so the correct solution is $q = \tfrac{1}{2}$.

Problem solved!

The equation must be in the form $ax^2 + bx + c = 0$ before you work out the values of a, b and c. Always write down the condition for the discriminant that you are using, and use **brackets** when you substitute.

> You will need to use problem-solving skills throughout your exam – **be prepared!**

Worked example

The equation $2x^2 - kx + 6 = k$ has no real solutions for x. Show that $k^2 + 8k - 48 < 0$
   (3 marks)

$$2x^2 - kx + 6 - k = 0$$
$$b^2 - 4ac < 0$$
$$(-k)^2 - 4 \times 2 \times (6 - k) < 0$$
$$k^2 + 8k - 48 < 0$$

Now try this

1   Find the value of the discriminant of
$3x^2 - 2x - 5$ **(1 mark)**

2   The equation $px^2 + 2x - 3 = 0$, where p is a constant, has equal roots.
Find the value of p.   **(3 marks)**

> The expression $(k + p)^2$ must always be greater than or equal to zero.

3   $f(x) = x^2 + (k + 5)x + 2k$, where k is a constant.

(a) Find the discriminant of $f(x)$ in terms of k.   **(2 marks)**

(b) Show that the discriminant can be written in the form $(k + p)^2 + q$, where p and q are integers to be found.   **(2 marks)**

(c) Show that, for all values of k, the equation $f(x) = 0$ has distinct real roots.   **(2 marks)**

Modelling with quadratics

You can use quadratic functions and graphs to **model** real life situations.

Projectiles

An object moving freely under gravity is called a projectile. If you ignore air resistance, the paths of projectiles can be modelled as quadratic curves. This model describes the path of an object projected from the top of a 20 m building:

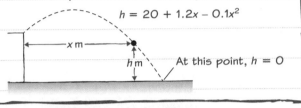

$$h = 20 + 1.2x - 0.1x^2$$

At this point, $h = 0$

Problem solved!

The constant b in this model tells you how much N will change for every unit change in P. You are told that the change in N will be -400 when P increases by 200. Use this information to write an equation and solve it to find b. Then use your value of b together with the fact that $N = 50\,000$ when $P = 17\,000$ to find a.

> You will need to use problem-solving skills throughout your exam – **be prepared!**

You need to complete the square for the expression $P(84\,000 - 2P)$. Start by multiplying out the brackets. Then take out a factor of -2 so the coefficient of P^2 is 1.

Your answer should refer to the **context of the question**. You should check your answer makes sense. £21 000 sounds about right for the price of a new car.

Worked example

A car manufacturer currently sells 50 000 units of a popular model each year, at a recommended retail price of £17 000 each. The manufacturer determines that for every £200 they increase the price, they will sell 400 fewer cars each year.

The number of cars sold each year, N, for a given price £P, is modelled as $N = a + bP$

(a) Determine the values of the constants a and b in this model.   **(2 marks)**

$$200b = -400$$
$$b = -2$$
$$50\,000 = a - 2(17\,000)$$
$$a = 84\,000$$

The model is $N = 84\,000 - 2P$

The total revenue generated, £X, can be modelled as $X = P(a - bP)$, where a and b are the same constants.

(b) Rearrange X into the form $c - d(P - e)^2$, where c, d and e are constants to be found.   **(3 marks)**

$$X = P(84\,000 - 2P)$$
$$= -2P^2 + 84\,000P$$
$$= -2(P^2 - 42\,000P)$$
$$= -2[(P - 21\,000)^2 - 21\,000^2]$$
$$= 441\,000\,000 - 2(P - 21\,000)^2$$

(c) State, with reasons, the amount of money the manufacturer should charge to maximise their revenue, and write down the maximum revenue.   **(2 marks)**

The revenue will be maximised when $2(P - 21\,000)^2$ takes its minimum value (zero). This occurs when $P = 21\,000$, so the manufacturer should charge £21 000. This will generate a revenue of £441 million.

Now try this

A stone is thrown from a clifftop at an angle of 35° above the horizontal. The height, h metres, of the stone above the water when it is a horizontal distance d metres from the cliff is modelled by the formula:

$$h = 42 + 0.7d - 0.14d^2, \; d \geq 0$$

(a) Write down the height of the clifftop.   **(1 mark)**

(b) Determine the horizontal distance travelled by the stone at the time it hits the water.   **(3 marks)**

(c) By completing the square, or otherwise, determine the maximum height of the stone above the water during its flight.   **(4 marks)**

Simultaneous equations

If a pair of simultaneous equations involves an x^2 or a y^2 term, you need to solve them using **substitution**. Remember to **number** the equations to keep track of your working.

Rearrange the linear equation to make y the subject.

$$y = x^2 - 2x - 7 \qquad ①$$
$$x - y = -3 \qquad ②$$

From ②:

$$y = x + 3 \qquad ③$$

Each solution for x has a corresponding value of y. Substitute to find the values of y.

$$x + 3 = x^2 - 2x - 7 \qquad \text{Substitute } x + 3 \text{ for } y \text{ in equation } ①.$$
$$0 = x^2 - 3x - 10$$
$$0 = (x - 5)(x + 2)$$
$$x = 5 \text{ or } x = -2$$

The solutions are $x = 5$, $y = 8$ and $x = -2$, $y = 1$.

Worked example

Solve the simultaneous equations

$$x - 2y = 1 \qquad ①$$
$$x^2 + y^2 = 13 \qquad ②$$

(6 marks)

From ①: $x = 1 + 2y$ ③

Substitute $1 + 2y$ for x in ②:

$$(1 + 2y)^2 + y^2 = 13$$
$$1 + 4y + 4y^2 + y^2 = 13$$
$$5y^2 + 4y - 12 = 0$$
$$(5y - 6)(y + 2) = 0$$

$$y = \frac{6}{5} \qquad \text{or} \qquad y = -2$$

$$x = 1 + 2\left(\frac{6}{5}\right) = \frac{17}{5} \qquad x = 1 + 2(-2) = -3$$

Solutions: $x = \frac{17}{5}$, $y = \frac{6}{5}$ and $x = -3$, $y = -2$

You can substitute for x or y. It is easier to substitute for x because there will be no fractions.

Use brackets to make sure that the whole expression is squared.

Rearrange the quadratic equation for y into the form $ay^2 + by + c = 0$

Factorise the left-hand side to find two solutions for y.

Remember that there will be **two pairs** of solutions. Each value of y will produce a corresponding value of x. You need to find **four** different values in total **and** pair them up correctly.

Thinking graphically

The solutions to a pair of simultaneous equations correspond to the points where the graphs of the equations **intersect**.
Because an equation involving x^2 or y^2 represents a **curve**, there can be more than one point of intersection. Each point has an x-value and a y-value. You can write the solutions using coordinates.

There is more on intersections of graphs on page 16.

$y = (1 + x)^2$

$y = x + 1$

$(0, 1)$

$(-1, 0)$

Now try this

1 Solve the simultaneous equations
$$x + y = 5$$
$$x^2 + 2y^2 = 22 \qquad \textbf{(6 marks)}$$

You will need to write $x^2 - 6x + 7$ in completed square form. Have a look at page 4 for a reminder.

2 (a) By eliminating y from the simultaneous equations
$$y = x + 6$$
$$xy - 2x^2 = 7$$
show that $x^2 - 6x + 7 = 0$ **(2 marks)**

(b) Hence solve the simultaneous equations in part (a), giving your answers in the form $a \pm \sqrt{2}$, where a is an integer.   **(4 marks)**

Inequalities

You might need to find a set of values which **satisfy** an inequality. If the inequality involves a **quadratic** expression you should always **sketch a graph** to help you answer the question.

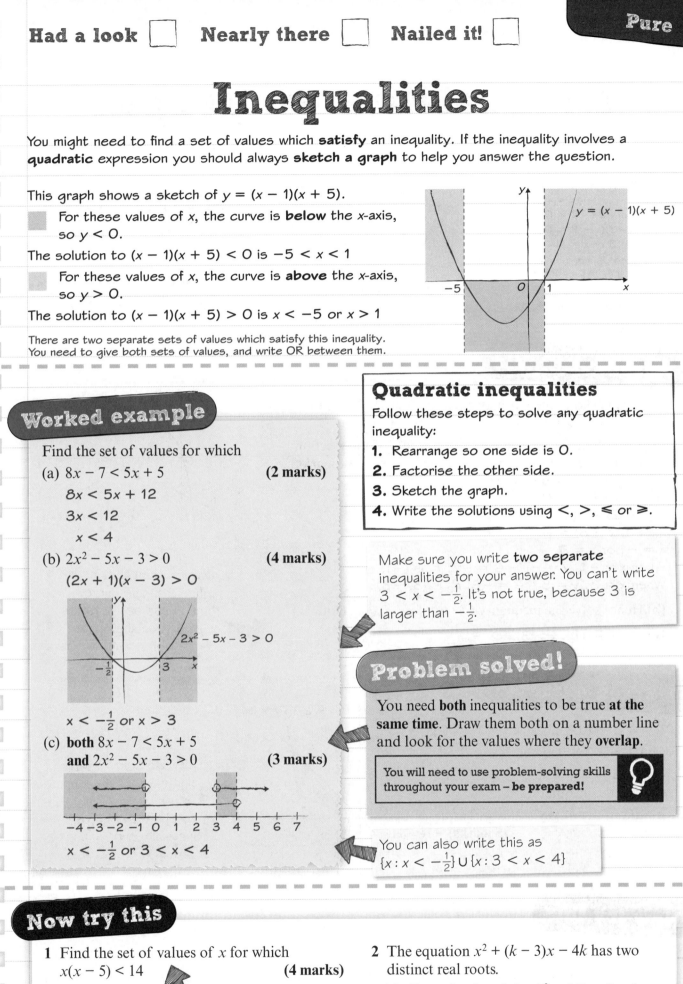

This graph shows a sketch of $y = (x - 1)(x + 5)$.

▨ For these values of x, the curve is **below** the x-axis, so $y < 0$.

The solution to $(x - 1)(x + 5) < 0$ is $-5 < x < 1$

▨ For these values of x, the curve is **above** the x-axis, so $y > 0$.

The solution to $(x - 1)(x + 5) > 0$ is $x < -5$ or $x > 1$

There are two separate sets of values which satisfy this inequality. You need to give both sets of values, and write OR between them.

$y = (x - 1)(x + 5)$

Worked example

Find the set of values for which

(a) $8x - 7 < 5x + 5$ **(2 marks)**

$8x < 5x + 12$

$3x < 12$

$x < 4$

(b) $2x^2 - 5x - 3 > 0$ **(4 marks)**

$(2x + 1)(x - 3) > 0$

$2x^2 - 5x - 3 > 0$

$x < -\frac{1}{2}$ or $x > 3$

(c) **both** $8x - 7 < 5x + 5$
and $2x^2 - 5x - 3 > 0$ **(3 marks)**

$-4 \ -3 \ -2 \ -1 \ \ 0 \ \ 1 \ \ 2 \ \ 3 \ \ 4 \ \ 5 \ \ 6 \ \ 7$

$x < -\frac{1}{2}$ or $3 < x < 4$

Quadratic inequalities

Follow these steps to solve any quadratic inequality:

1. Rearrange so one side is 0.
2. Factorise the other side.
3. Sketch the graph.
4. Write the solutions using $<$, $>$, $\leqslant$ or $\geqslant$.

Make sure you write **two separate** inequalities for your answer. You can't write $3 < x < -\frac{1}{2}$. It's not true, because 3 is larger than $-\frac{1}{2}$.

Problem solved!

You need **both** inequalities to be true **at the same time**. Draw them both on a number line and look for the values where they **overlap**.

> You will need to use problem-solving skills throughout your exam – **be prepared!**

You can also write this as $\{x : x < -\frac{1}{2}\} \cup \{x : 3 < x < 4\}$

Now try this

1 Find the set of values of x for which
$x(x - 5) < 14$ **(4 marks)**

> You need to expand the brackets and rearrange the inequality into the form $ax^2 + bx + c < 0$ first.

2 The equation $x^2 + (k - 3)x - 4k$ has two distinct real roots.

(a) Show that k satisfies $k^2 + 10k + 9 > 0$ **(3 marks)**

(b) Hence find the set of possible values for k.   **(4 marks)**

Inequalities on graphs

You can interpret inequalities graphically by considering curves and regions. This diagram shows the graphs of $y = x^2 - 4x$ and $y = -2x + 3$:

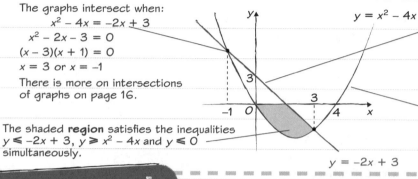

The graphs intersect when:
$$x^2 - 4x = -2x + 3$$
$$x^2 - 2x - 3 = 0$$
$$(x - 3)(x + 1) = 0$$
$$x = 3 \text{ or } x = -1$$

There is more on intersections of graphs on page 16.

The shaded **region** satisfies the inequalities $y \leqslant -2x + 3$, $y \geqslant x^2 - 4x$ and $y \leqslant 0$ simultaneously.

Between the points of intersection the line is **above** the curve. So the solution to $-2x + 3 > x^2 - 4x$ is $-1 < x < 3$.

Outside the points of intersection the line is **below** the curve. So the solution to $-2x + 3 < x^2 - 4x$ is $x < -1$ or $x > 3$.

Worked example

The graph shows the line with equation $y = x - 6$ and the curve with equation $y = 5x - 2x^2$.

(a) Determine the coordinates of the points of intersection of the line and the curve. **(4 marks)**

$$x - 6 = 5x - 2x^2$$
$$2x^2 - 4x - 6 = 0$$
$$x^2 - 2x - 3 = 0$$
$$(x + 1)(x - 3) = 0$$
$$x = -1 \text{ or } x = 3$$

When $x = -1$, $y = -1 - 6 = -7$. P_1 is $(-1, -7)$
When $x = 3$, $y = 3 - 6 = -3$. P_2 is $(3, -3)$

(b) Hence solve the inequality $x - 6 > 5x - 2x^2$. **(1 mark)**

$\{x : x < -1\} \cup \{x : x > 3\}$

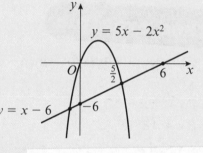

The solutions are the x-values when the line is higher than the curve. You can use set notation or give your answer as '$x < -1$ or $x > 3$'.

Worked example

$f(x) = x^2 - 6x + 5$

$g(x) = x$

On a graph, show the region that satisfies the inequalities:

$y > f(x)$ $y \leqslant g(x)$ $x > 2$

(5 marks)

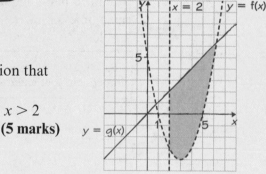

Use dotted lines to show a strict inequality ($<$ or $>$) and solid lines to show a non-strict inequality ($\leqslant$ or $\geqslant$).

Now try this

$f(x) = 6 - x^2$ $g(x) = \frac{1}{2}x + 1$

(a) On the same axes, sketch the graphs of $y = f(x)$ and $y = g(x)$. Indicate any points of intersection with the coordinate axes. **(3 marks)**

(b) Determine the x-coordinates of any points of intersection of the two graphs. **(3 marks)**

(c) Hence solve the inequalities
   (i) $f(x) < g(x)$ (ii) $f(x) \geqslant g(x)$ **(2 marks)**

(d) Shade the region on the graph that satisfies all of the following inequalities:
   $y \leqslant f(x)$, $y \geqslant g(x)$, $x \geqslant 0$ **(1 mark)**

Cubic and quartic graphs

In a cubic function, the highest power of x is x^3. In a quartic function, it is x^4. You need to know the shapes of graphs of cubic and quartic functions and be able to sketch them.

Factorise then sketch

You can sketch the graphs of cubic and quartic functions by **factorising** them to find their roots.

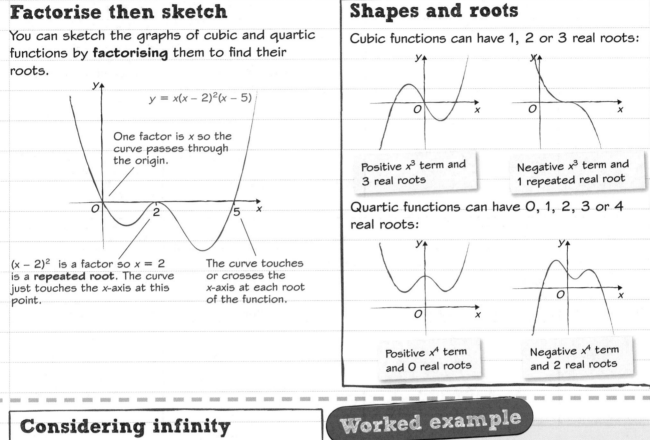

$y = x(x - 2)^2(x - 5)$

One factor is x so the curve passes through the origin.

$(x - 2)^2$ is a factor so $x = 2$ is a **repeated root**. The curve just touches the x-axis at this point.

The curve touches or crosses the x-axis at each root of the function.

Shapes and roots

Cubic functions can have 1, 2 or 3 real roots:

Positive x^3 term and 3 real roots

Negative x^3 term and 1 repeated real root

Quartic functions can have 0, 1, 2, 3 or 4 real roots:

Positive x^4 term and 0 real roots

Negative x^4 term and 2 real roots

Considering infinity

In the example on the right, as x gets large, the x^3 term gets large **more quickly** than the x^2 term. So for large positive x, y gets very large. You can write 'as $x \to \infty$, $y \to \infty$'.

Similarly, as $x \to -\infty$, $y \to -\infty$.

This tells you how the curve will behave at either end of the x-axis.

The factorised equation has a factor of x so the curve will pass through the origin. It also has a **repeated** factor of $(x - 4)$ so the curve will just touch the x-axis at the point $x = 4$.

Worked example

(a) Factorise completely $x^3 - 8x^2 + 16x$ **(3 marks)**

$x(x^2 - 8x + 16) = x(x - 4)^2$

(b) Hence sketch the curve with equation $y = x^3 - 8x^2 + 16x$, showing the points where the curve meets the coordinate axes. **(3 marks)**

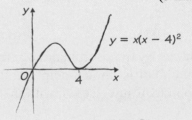

$y = x(x - 4)^2$

Now try this

1 (a) Factorise completely $x^3 - 9x$ **(3 marks)**

(b) Hence sketch the curve $y = x^3 - 9x$ **(3 marks)**

2 Sketch the graph of $y = (2x - 1)(x - 3)^2$, showing clearly the coordinates of the points where the curve meets the coordinate axes. **(4 marks)**

3 Sketch the graph of $y = x(5 - x)(2x^2 + 9x + 4)$. Show clearly the coordinates of any points where the curve meets or crosses the coordinate axes. **(4 marks)**

You need to show the coordinates of the point where the graph meets the y-axis as well.

Transformations 1

You can change the equation of a graph to translate it, stretch it or reflect it. These tables show you how you can use functions to transform the graph of $y = f(x)$.

Function	$y = f(x) + a$	$y = f(x + a)$	$y = af(x)$
Transformation of graph	Translation $\begin{pmatrix} 0 \\ a \end{pmatrix}$	Translation $\begin{pmatrix} -a \\ 0 \end{pmatrix}$	Stretch in the vertical direction, scale factor a
Useful to know	$f(x) + a \rightarrow$ move UP a units $f(x) - a \rightarrow$ move DOWN a units	$f(x + a) \rightarrow$ move LEFT a units $f(x - a) \rightarrow$ move RIGHT a units	x-values stay the same
Example	$y = f(x) + 3$ $y = f(x)$	$y = f(x)$ $y = f(x + 5)$	$y = 3f(x)$ $y = f(x)$

Function	$y = f(ax)$	$y = -f(x)$	$y = f(-x)$
Transformation of graph	Stretch in the horizontal direction, scale factor $\frac{1}{a}$	Reflection in the x-axis	Reflection in the y-axis
Useful to know	y-values stay the same	'$-$' outside the bracket	'$-$' inside the bracket
Example	$y = f(2x)$ $y = f(x)$	$y = f(x)$ $y = -f(x)$	$y = f(-x)$ $y = f(x)$

Worked example

The diagram shows a sketch of a curve with equation $y = f(x)$.

On the same diagram sketch the curve with equation
(a) $y = f(x + 3)$ **(3 marks)**
(b) $y = -f(x)$. **(3 marks)**

Show clearly the coordinates of any maximum or minimum points, and any points of intersection with the axes.

Everything in blue is part of the answer.

$y = f(x + 3)$ $y = -f(x)$ $(3, 2)$ -3 2 5 $(3, -2)$ $y = f(x)$

Now try this

The diagram shows a sketch of a curve C with equation $y = f(x)$.
On separate diagrams sketch the curve with equation
(a) $y = 2f(x)$ **(3 marks)**   (b) $y = f(-x)$ **(3 marks)**
(c) $y = f(x + k)$, where k is a constant and $0 < k < 4$ **(4 marks)**

On each diagram show the coordinates of any maximum or minimum points, and any points of intersection with the x-axis.

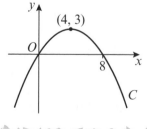

$(4, 3)$ 8 C

Had a look ☐ Nearly there ☐ Nailed it! ☐

Transformations 2

You need to be able to spot transformed functions from their equations, and sketch transformations involving **asymptotes**.

Functions and equations

Curve C below has equation $y = x^3 - 3x^2 - 2$. You can sketch the curves of other equations by transforming curve C.

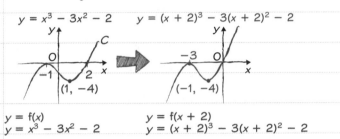

$y = x^3 - 3x^2 - 2$ $y = (x + 2)^3 - 3(x + 2)^2 - 2$

$y = f(x)$
$y = x^3 - 3x^2 - 2$

$y = f(x + 2)$
$y = (x + 2)^3 - 3(x + 2)^2 - 2$

Asymptotes

An asymptote is a line which a curve approaches, but never reaches. You draw asymptotes on graphs with DOTTED LINES.

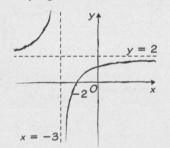

This curve has an asymptote at $y = 5$. When you transform a graph, its asymptotes are transformed as well.

Transformation	New asymptote
$y = f(x) - 1$	$y = 4$
$y = 2f(x)$	$y = 10$
$y = f(x + 4)$	$y = 5$

The graph is translated 4 units to the left, so the horizontal asymptote does not change.

The diagram shows a sketch of the curve with equation $y = f(x)$ where

$$f(x) = \frac{2x}{x + 1}, \; x \neq -1$$

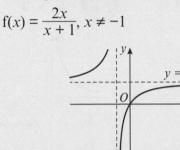

The curve has asymptotes with equations $y = 2$ and $x = -1$

(a) Sketch the curve with equation $y = f(x + 2)$ and state the equations of its asymptotes. **(3 marks)**

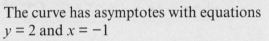

(b) Find the coordinates of the points where the curve in part (a) crosses the coordinate axes. **(3 marks)**

$$f(x + 2) = \frac{2(x + 2)}{(x + 2) + 1} = \frac{2x + 4}{x + 3}$$

When $x = 0$, $y = \frac{4}{3}$

$(-2, 0)$ and $(0, \frac{4}{3})$

Now try this

The diagram shows a curve C with equation $y = f(x)$, where

$$f(x) = \frac{(x + 2)^2}{x + 1}, \; x \neq -1$$

(a) Sketch the curve with equation $y = f(x + 1)$ and state the new equation of the asymptote $x = -1$ **(3 marks)**

(b) Write down the coordinates of the points where the curve meets the coordinate axes. **(3 marks)**

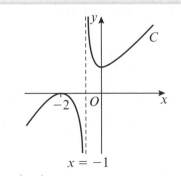

Reciprocal graphs

You need to know how to sketch the graphs of $y = \dfrac{k}{x}$ and $y = \dfrac{k}{x^2}$, and transformations of these graphs.

Shapes and asymptotes

The shapes of the reciprocal graphs are different for **positive** and **negative** values of k:

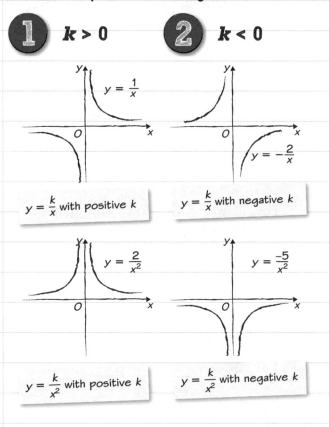

① $k > 0$ ② $k < 0$

$y = \dfrac{1}{x}$

$y = -\dfrac{2}{x}$

$y = \dfrac{k}{x}$ with positive k $y = \dfrac{k}{x}$ with negative k

$y = \dfrac{2}{x^2}$

$y = \dfrac{-5}{x^2}$

$y = \dfrac{k}{x^2}$ with positive k $y = \dfrac{k}{x^2}$ with negative k

The graphs of $y = \dfrac{k}{x}$ and $y = \dfrac{k}{x^2}$ have **asymptotes** at the x-axis and the y-axis. Remember to translate any asymptotes when you translate the graph.

Worked example

The figure shows a sketch of the curve $y = \dfrac{6}{x}$

$y = \dfrac{6}{x}$

(a) On a separate diagram, sketch the curve with equation $y = \dfrac{6}{x - 3}$, showing any points at which the curve crosses the coordinate axes.   **(3 marks)**

When $x = 0$, $y = \dfrac{6}{-3} = -2$

(b) Write down the equation of the asymptotes of the curve in part (a).   **(2 marks)**

$y = 0$ and $x = 3$

The transformation from $y = \dfrac{6}{x}$ to $y = \dfrac{6}{x - 3}$ is the translation $\begin{pmatrix} 3 \\ 0 \end{pmatrix}$.

Now try this

1 Sketch the graph of $y = -\dfrac{4}{x}$ **(2 marks)**

The transformation is
$y = f(x) \rightarrow y = f(x + 1)$
Draw the new asymptote on your sketch before you draw your curve.

2 (a) Sketch the graph of $y = \dfrac{3}{x}$ **(2 marks)**

(b) On a separate diagram, sketch the graph of $y = \dfrac{3}{x + 1}$, showing any points at which the curve crosses the coordinate axes.   **(3 marks)**

(c) Write down the equations of the asymptotes of the curve in part (b).   **(2 marks)**

Points of intersection

The coordinates of the points where two graphs **intersect** are the x- and y-values which satisfy **both** equations at the same time. You can use algebra to find the points where two curves intersect.

The diagram shows the graphs of $y = -\dfrac{2}{x}$ and $y = 5 - 3x$.

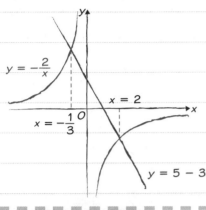

The x-coordinates at the points of intersection are the solutions to the equation

$$5 - 3x = -\frac{2}{x}$$
$$x(5 - 3x) = -2$$
$$5x - 3x^2 = -2$$
$$3x^2 - 5x - 2 = 0$$
$$(3x + 1)(x - 2) = 0$$
$$x = -\frac{1}{3} \quad \text{or} \quad x = 2$$

Worked example

(a) On the same axes, sketch the graph with the equation
   (i) $y = x(x + 1)(x - 4)$
   (ii) $y = \dfrac{2}{x}$ **(5 marks)**

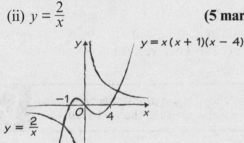

(b) Write down the number of real solutions to the equation $x(x + 1)(x - 4) = \dfrac{2}{x}$

(1 mark)

2

The points of intersection will be solutions to the equation $x^2(3 - x) = -4x$.

Worked example

(a) On the same axes, sketch the graphs with the equations
   (i) $y = x^2(3 - x)$
   (ii) $y = -4x$ **(5 marks)**

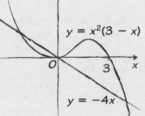

(b) Find the coordinates of the points of intersection.   **(6 marks)**

$$3x^2 - x^3 = -4x$$
$$x^3 - 3x^2 - 4x = 0$$
$$x(x^2 - 3x - 4) = 0$$
$$x(x - 4)(x + 1) = 0$$
$$x = 0 \quad \text{or} \quad x = 4 \quad \text{or} \quad x = -1$$
$$y = 0 \qquad y = -16 \qquad y = 4$$
$$(0, 0), \ (4, -16), \ (-1, 4)$$

Now try this

(a) On the same axes, sketch the graph with the equation
   (i) $y = x^2(x - 3)$
   (ii) $y = x(8 - x)$ **(6 marks)**
   Indicate all the points where the curves meet the x-axis.

(b) Use algebra to find the coordinates of the points of intersection.   **(7 marks)**

There are three points of intersection: one at (0, 0), one with a negative value of x and one with a positive value of x.

16

Equations of lines

The equation of a straight line can be written in the form $y = mx + c$, where m is the **gradient** of the line, and c is the point where it crosses the y-axis. There are other useful ways to write the equation of a straight line.

Point and gradient

If a straight line has **gradient** m and passes through the **point** (x_1, y_1), then you can write its equation as

$$y - y_1 = m(x - x_1)$$

This equation is **very** useful for lots of exam questions! Make sure you learn it.

If you are given two points on a straight line, (x_1, y_1) and (x_2, y_2), you can calculate the gradient using

$$m = \frac{y_2 - y_1}{x_2 - x_1}$$

Thinking in transformations

You can remember this equation by thinking of it as a **translation** of the graph $y = mx$ by vector $\begin{pmatrix} x_1 \\ y_1 \end{pmatrix}$

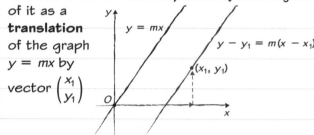

Worked example

The line L passes through the point $(-8, 5)$ and has gradient $\frac{1}{2}$. Find an equation for L in the form $ax + by + c = 0$, where a, b and c are integers. **(3 marks)**

$$y - y_1 = m(x - x_1)$$

$$y - 5 = \tfrac{1}{2}(x - (-8))$$

$$2y - 10 = x + 8$$

$$x - 2y + 18 = 0$$

Problem solved!

If you are using a formula which is **not** in the booklet, always **write it down** before you substitute. Here, $m = \frac{1}{2}$, $x_1 = -8$ and $y_1 = 5$.

You need a, b and c to be integers, so multiply every term in your equation by 2 to remove the fraction. Then rearrange so one side is equal to 0.

You will need to use problem-solving skills throughout your exam – **be prepared!**

You can draw a sketch to help you find the gradient, or use

$$m = \frac{y_2 - y_1}{x_2 - x_1}$$

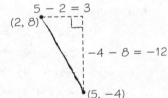

Write the equation as $y = -4x + c$ and substitute the values of x and y from either point on the graph to find c.

Worked example

The line L passes through the points $(2, 8)$ and $(5, -4)$. Find an equation for L in the form $y = mx + c$ **(3 marks)**

$$m = \frac{y_2 - y_1}{x_2 - x_1} = \frac{-4 - 8}{5 - 2} = \frac{-12}{3} = -4$$

$$y = -4x + c$$

$$8 = -4(2) + c$$

$$c = 16 \quad \text{so} \quad y = -4x + 16$$

Now try this

1 The line L passes through the point $(6, -5)$ and has gradient $-\frac{1}{3}$. Find an equation for L in the form $ax + by + c = 0$, where a, b and c are integers. **(3 marks)**

2 The line L passes through $(-4, 2)$ and $(8, 11)$. Find an equation for L in the form $y = mx + c$, where m and c are constants. **(3 marks)**

3 The line $3y + 4x - k = 0$ passes through the point $(5, 1)$. Find
   (a) the value of k **(1 mark)**
   (b) the gradient of the line. **(2 marks)**

To find the gradient, rearrange the equation into the form $y = mx + c$.

Parallel and perpendicular

Parallel lines have the same gradient. These three lines all have a gradient of 1.

Perpendicular means at right angles. If a line has gradient m then any line perpendicular to it will have gradient $-\dfrac{1}{m}$

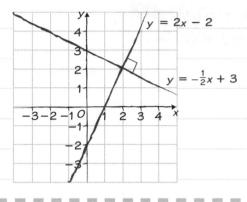

Worked example

P is the point $(3, -2)$ and Q is the point $(7, 6)$. The line L is perpendicular to PQ and passes through the midpoint of PQ. Find an equation for L in the form $ax + by + c = 0$, where a, b and c are integers. **(5 marks)**

Gradient of $PQ = \dfrac{6 - (-2)}{7 - 3} = \dfrac{8}{4} = 2$

So gradient of $L = -\dfrac{1}{2}$

Midpoint of $PQ = \left(\dfrac{3 + 7}{2}, \dfrac{-2 + 6}{2}\right)$

$= (5, 2)$

Equation of L: $y - y_1 = m(x - x_1)$

$y - 2 = -\dfrac{1}{2}(x - 5)$

$2y - 4 = -x + 5$

$x + 2y - 9 = 0$

Worked example

The line L_1 has equation $7y + 2x - 3 = 0$. The line L_2 is perpendicular to L_1 and crosses the y-axis at $(0, 5)$. Find an equation for L_2 in the form $ax + by + c = 0$, where a, b and c are integers. **(3 marks)**

L_1: $7y = -2x + 3$

$y = -\dfrac{2}{7}x + \dfrac{3}{7}$

L_2 has gradient $\dfrac{7}{2}$

L_2: $y = \dfrac{7}{2}x + 5$

$2y = 7x + 10$

$7x - 2y + 10 = 0$

> Start by finding the gradient of L_1. The easiest way to do this is to rearrange the equation into the form $y = mx + c$.

> Find the gradient and the midpoint of PQ. The midpoint of the line segment joining (x_1, y_1) and (x_2, y_2) is $\left(\dfrac{x_1 + x_2}{2}, \dfrac{y_1 + y_2}{2}\right)$.

Now try this

1 The line L has equation $y = 10 - 3x$

  (a) Show that the point $P (4, -2)$ lies on L. **(1 mark)**

  (b) Find an equation of the line perpendicular to L which passes through P. Give your answer in the form $ax + by + c = 0$, where a, b and c are integers. **(3 marks)**

2 The line L_1 with equation $4x - 5y - 1 = 0$ crosses the x-axis at A. The line L_2 is perpendicular to L_2 and passes through A. Find the equation of L_2 in the form $y = mx + c$ **(4 marks)**

> Substitute $y = 0$ into the equation of L_1 to work out the x-coordinate of A.

Lengths and areas

You might have to calculate the length of a line segment, or the area of a shape on a coordinate grid. It's always a good idea to **draw sketches** to keep track of your working.

Worked example

P is the point $(2, 5)$ and Q is the point $(3, -2)$. The length of PQ is $a\sqrt{2}$, where a is an integer. Find the value of a.

(3 marks)

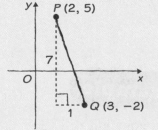

$PQ^2 = 1^2 + 7^2$

$\quad = 50$

$PQ = \sqrt{50}$

$\quad = \sqrt{25} \times \sqrt{2} = 5\sqrt{2}$

$a = 5$

Using a formula

If P has coordinates (x_1, y_1) and Q has coordinates (x_2, y_2), then the length of the line segment PQ is

$$\sqrt{(x_2 - x_1)^2 + (y_2 - y_1)^2}$$

(a) To work out the coordinates of the point where two lines intersect you need to solve their equations **simultaneously**.

Substitute $y = -x$ into the equation for L_2.

(b) The vertical height of triangle AOB is $\frac{5}{2}$. Substitute $y = 0$ into the equation for L_2 to find the coordinates of B, then use Area $= \frac{1}{2} \times$ base $\times$ height. You don't need to give any units when you're calculating lengths and areas on a graph.

Worked example

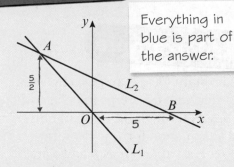

Everything in blue is part of the answer.

The line L_1 has equation $y = -x$, and the line L_2 has equation $3y + x - 5 = 0$

(a) Work out the coordinates of A. **(2 marks)**

$3(-x) + x - 5 = 0$

$\quad -2x - 5 = 0$

$\qquad x = -\frac{5}{2}, \ y = \frac{5}{2}$

A is the point $\left(-\frac{5}{2}, \frac{5}{2}\right)$

(b) Find the area of triangle AOB. **(3 marks)**

$3(0) + x - 5 = 0$

$\qquad x = 5$

B is the point $(5, 0)$.

Area $= \frac{1}{2} \times 5 \times \frac{5}{2} = \frac{25}{4}$

Now try this

The line L_1 has equation $x - 2y + 6 = 0$
L_1 crosses the x-axis at P and the y-axis at Q.

(a) Show that $PQ = 3\sqrt{5}$ **(3 marks)**

The line L_2 is perpendicular to L_1 and passes through Q.

(b) Find an equation for L_2. **(4 marks)**

L_2 crosses the x-axis at R.

(c) Find the area of the triangle PQR. **(4 marks)**

Draw a sketch to help you visualise the problem.

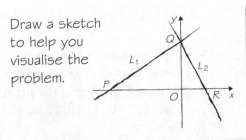

Equation of a circle

A circle with centre (a, b) and radius r has equation

$$(x - a)^2 + (y - b)^2 = r^2$$

Be really careful with the right-hand side. It is the radius **squared**.

The formulae booklet does not contain **any** coordinate geometry formulae! You will need to learn them for your exam.

$(x - 2)^2 + (y - 4)^2 = 9$

$r = 3$

$(2, 4)$

Worked example

A circle C has centre $(3, 1)$ and passes through the point $(-2, 5)$.

(a) Find an equation for C. **(4 marks)**

$r = \sqrt{(x_2 - x_1)^2 + (y_2 - y_1)^2}$

$\quad = \sqrt{(-2 - 3)^2 + (5 - 1)^2} = \sqrt{41}$

$(x - a)^2 + (y - b)^2 = r^2$

$(x - 3)^2 + (y - 1)^2 = 41$

(b) Verify that $(7, -4)$ also lies on C. **(1 mark)**

$(7 - 3)^2 + (-4 - 1)^2 = 4^2 + (-5)^2 = 41$

So point $(7, -4)$ lies on C.

The radius of the circle, **r**, is the length of the line segment between $(3, 1)$ and $(-2, 5)$. For a reminder about finding the length of a line segment have a look at page 19.

Just because you can use a calculator in your exam, it doesn't mean you always should!

Don't write your radius as a decimal. If you leave it in the form $\sqrt{41}$ it will be exact when you square it to write your equation.

Problem solved!

You need to rearrange the equation into the form $(x - a)^2 + (x - b)^2 = r^2$

This is a bit like **completing the square**. Have a look at page 4 for a recap.

$x^2 - 6x = (x - 3)^2 - 3^2$

$y^2 + 2y = (y + 1)^2 - 1^2$

Remember that the right-hand side of the equation is r^2, so the radius is 5, not 25.

> You will need to use problem-solving skills throughout your exam – **be prepared!**

Substitute $x = 0$ into the equation to find the points where C crosses the y-axis.

Worked example

The circle C has equation $x^2 + y^2 - 6x + 2y - 15 = 0$

(a) Find the coordinates of the centre of C and the radius of C. **(4 marks)**

$x^2 - 6x + y^2 + 2y - 15 = 0$

$(x - 3)^2 - 3^2 + (y + 1)^2 - 1^2 - 15 = 0$

$(x - 3)^2 + (y + 1)^2 - 25 = 0$

$(x - 3)^2 + (y + 1)^2 = 5^2$

C has centre $(3, -1)$ and radius 5.

(b) Find the coordinates of the points where C crosses the y-axis. **(2 marks)**

When $x = 0$, $(0 - 3)^2 + (y + 1)^2 = 25$

$(y + 1)^2 = 16$

$y + 1 = \pm 4$

$y = -5$ or 3

C crosses the y-axis at $(0, -5)$ and $(0, 3)$.

Now try this

$A(-6, 0)$ and $B(2, 4)$ are the endpoints of a diameter of the circle C. Find

(a) the length of AB **(2 marks)**

(b) the coordinates of the midpoint of AB **(2 marks)**

(c) an equation for the circle C. **(2 marks)**

You will need these two key **diameter** facts to solve this problem:

1. The **midpoint** of any diameter of a circle is the centre of the circle.

2. The diameter is **twice** the radius.

Have a look at page 18 for a reminder about midpoints.

Circle properties

You might need to solve circle problems involving **semicircles**, **tangents** and **chords**. Here are the three key facts you might need to use.

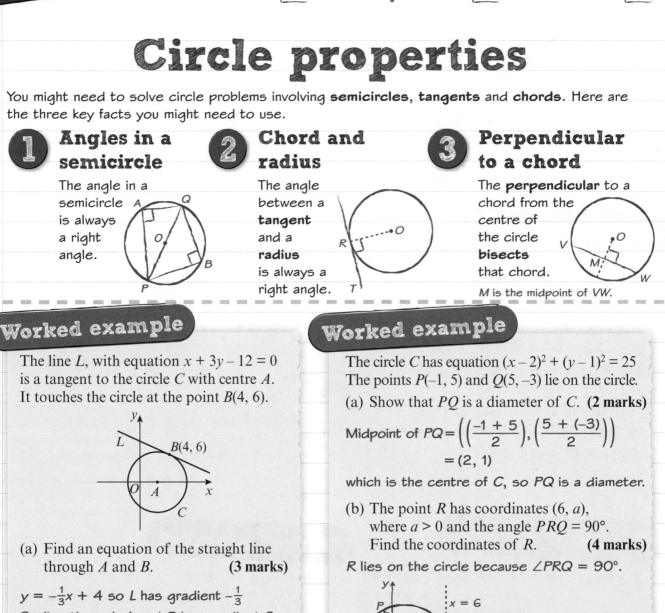

1 **Angles in a semicircle**

The angle in a semicircle is always a right angle.

2 **Chord and radius**

The angle between a **tangent** and a **radius** is always a right angle.

3 **Perpendicular to a chord**

The **perpendicular** to a chord from the centre of the circle **bisects** that chord.

M is the midpoint of VW.

Worked example

The line L, with equation $x + 3y - 12 = 0$ is a tangent to the circle C with centre A. It touches the circle at the point $B(4, 6)$.

(a) Find an equation of the straight line through A and B. **(3 marks)**

$y = -\frac{1}{3}x + 4$ so L has gradient $-\frac{1}{3}$

So line through A and B has gradient 3.

$y - y_1 = m(x - x_1)$

$y - 6 = 3(x - 4)$

$y = 3x - 6$

(b) Given that A lies on the x-axis, find the coordinates of A. **(1 mark)**

At A, $y = 0$ so $0 = 3x - 6$ so $x = 2$

A has coordinates (2, 0).

Worked example

The circle C has equation $(x - 2)^2 + (y - 1)^2 = 25$
The points $P(-1, 5)$ and $Q(5, -3)$ lie on the circle.

(a) Show that PQ is a diameter of C. **(2 marks)**

Midpoint of $PQ = \left(\left(\frac{-1 + 5}{2}\right), \left(\frac{5 + (-3)}{2}\right)\right)$

$= (2, 1)$

which is the centre of C, so PQ is a diameter.

(b) The point R has coordinates $(6, a)$, where $a > 0$ and the angle $PRQ = 90°$. Find the coordinates of R. **(4 marks)**

R lies on the circle because $\angle PRQ = 90°$.

> It can help to draw a sketch.

$x = 6$ so $(6 - 2)^2 + (y - 1)^2 = 25$

$(y - 1)^2 = 9$

$y = -2$ or 4

a > 0 so R has coordinates (6, 4).

Problem solved!

A tangent is perpendicular to a radius, so the straight line through A and B is perpendicular to L. Remember that if a line has gradient m, then any line perpendicular to it will have gradient $-\frac{1}{m}$.

Look at page 18 for a recap.

> You will need to use problem-solving skills throughout your exam – **be prepared!**

Now try this

The circle C has centre $(5, -2)$ and radius 10
(a) Write down an equation for C. **(2 marks)**
(b) Verify that the point $(-1, 6)$ lies on C. **(1 mark)**
(c) Find an equation of the tangent to C at the point $(-1, 6)$, giving your answer in the form $ax + by + c = 0$, where a, b and c are integers. **(4 marks)**

Circles and lines

You can solve the equations of a circle and a straight line **simultaneously** to find the points of intersection. A straight line will intersect a circle once, twice or not at all. If a straight line just touches a circle then it is a **tangent** to the circle.

There is more about solving simultaneous equations on page 9.

Worked example

The circle C has centre the origin and radius 3. The straight line with equation $2x + y = k$, where k is a positive constant, is a tangent to C. Find the exact value of k. **(6 marks)**

$y = k - 2x$ ①

$x^2 + y^2 = 9$ ②

Substitute ① into ②:

$x^2 + (k - 2x)^2 = 9$

$x^2 + 4x^2 - 4kx + k^2 = 9$

$5x^2 - 4kx + k^2 - 9 = 0$

Using the discriminant:

$b^2 - 4ac = 0$

$(-4k)^2 - 4(5)(k^2 - 9) = 0$

$16k^2 - 20k^2 + 180 = 0$

$k^2 = 45$

$k = 3\sqrt{5}$

Problem solved!

You can use the **discriminant** to solve this problem. You need to:

1. Write out the equation of the circle.

2. Substitute $y = k - 2x$ into this to solve both equations simultaneously.

3. Find the discriminant of the resulting quadratic and set it equal to 0.

4. Solve to find the value of k. You are told that k is positive so ignore the negative square root of 45.

You will need to use problem-solving skills throughout your exam – **be prepared!**

Circumcircles

For any triangle, you can draw a unique circle which passes through all three vertices. This is called the circumcircle of the triangle

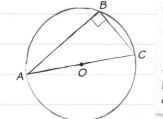

ABC is a right-angled triangle, so AC is a **diameter** of the circumcircle. Find the midpoint of the hypotenuse to find the centre of the circle.

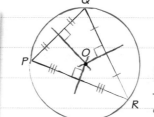

For any triangle, PQR, the **perpendicular bisectors** of the sides will intersect at the centre of the circumcircle.

There is more about properties of chords on page 21.

Now try this

1 Find the coordinates of the points where the line with equation $y = x + 7$ intersects the circle with equation $x^2 + (y + 2)^2 = 45$. **(5 marks)**

2 The line with equation $2x - y + 2 = 0$ intersects the circle with centre $(k, 0)$ and radius 2 at two distinct points. Find the range of possible values of k, giving your answer in surd form. **(7 marks)**

3 A circle passes through the points $(0, 0)$, $(0, 10)$ and $(8, -6)$, as shown in the diagram. Find an equation for the circle. **(7 marks)**

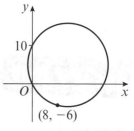

The perpendicular bisectors of two chords will intersect at the centre of the circle.

The factor theorem

In your exam, you might need to use the factor theorem to help you **factorise polynomials** like these:

$f(x) = 2x^3 + 5x^2 - 15x + 10$

This is a CUBIC because the highest power of x is 3.

$f(x) = x^4 - 4x^3 + 2x^2 - 3$

This is a QUARTIC because the highest power of x is 4.

The factor theorem

If f(x) is a polynomial and f(p) = 0, then (x – p) is a **factor** of f(x).

✓ Only use this theorem with **polynomials**.

✓ Watch out for the **sign**. If f(–1) = 0 then the factor would be (x + 1).

✓ **Learn** this rule – it's not in the formulae booklet.

Synthetic division with polynomials

If you have to **completely factorise** a **cubic** polynomial, you will usually need to find **three linear factors**. You can find the first one using the factor theorem. When you take this factor out, your other factor will be a **quadratic expression**. One quick way to find this is by using **synthetic division**.

To completely factorise $f(x) = 3x^3 - 17x^2 + 2x + 40$:
1. Use the factor theorem to find one factor: f(5) = 0, so (x – 5) is a factor.
2. Use synthetic division to divide $3x^3 - 17x^2 + 2x + 40$ by (x – 5):

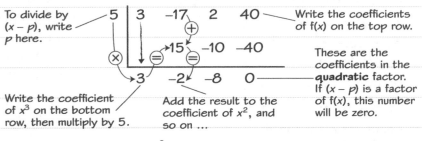

To divide by (x – p), write p here.

Write the coefficients of f(x) on the top row.

These are the coefficients in the **quadratic** factor. If (x – p) is a factor of f(x), this number will be zero.

Write the coefficient of x^3 on the bottom row, then multiply by 5.

Add the result to the coefficient of x^2, and so on ...

3. The quadratic factor is $3x^2 - 2x - 8$. Factorise this to complete the factorisation.
4. So $3x^3 - 17x^2 + 2x + 40 = (x - 5)(3x^2 - 2x - 8) = (x - 5)(3x + 4)(x - 2)$.

Worked example

$f(x) = 2x^3 - x^2 - 15x + 18$

(a) Use the factor theorem to show that (x + 3) is a factor of f(x). **(2 marks)**

$f(-3) = 2(-3)^3 - (-3)^2 - 15(-3) + 18$
$= -54 - 9 + 45 + 18 = 0$

So (x + 3) is a factor.

(b) Factorise f(x) completely. **(4 marks)**

Using synthetic division to divide f(x) by (x + 3):

```
-3 | 2   -1   -15   18
   |      -6    21  -18
   ----------------------
     2   -7     6    0
```

So $\dfrac{2x^3 - x^2 - 15x + 18}{x + 3} = 2x^2 - 7x + 6$

$f(x) = (x + 3)(2x^2 - 7x + 6) = (x + 3)(2x - 3)(x - 2)$

Be careful with the sign (+ or –). The factor theorem says that (x – p) is a factor if f(p) = 0, so you need to evaluate f(–3). You need to write down that (x + 3) is a factor at the end of your working.

You know that (x + 3) is a factor, so divide f(x) by (x + 3) using **synthetic division**. You could check your answer by expanding the brackets:

$(x + 3)(2x^2 - 7x + 6)$

$= 2x^3 - 7x^2 + 6x + 6x^2 - 21x + 18$
$= 2x^3 - x^2 - 15x + 18$ ✔

You need to factorise $(2x^2 - 7x + 6)$ in the normal way. Look at page 2 for a reminder.

Now try this

1 (a) Use the factor theorem to show that (x – 2) is a factor of $x^3 - 7x^2 - 14x + 48$ **(2 marks)**

(b) Factorise $x^3 - 7x^2 - 14x + 48$ completely. **(4 marks)**

2 $f(x) = 2x^3 - 3x^2 - 65x - a$

(a) Given that (x + 4) is a factor of f(x), find the value of a. **(3 marks)**

(b) Factorise f(x) completely. **(4 marks)**

The binomial expansion

The binomial expansion is a formula that lets you **expand brackets** easily. This is how the binomial expansion will appear in the formulae booklet in your AS exam.

$$(a + b)^n = a^n + \binom{n}{1}a^{n-1}b + \binom{n}{2}a^{n-2}b^2 + \ldots + \binom{n}{r}a^{n-r}b^r + \ldots + b^n \quad (n \in \mathbb{N})$$

where $\binom{n}{r} = {}^nC_r = \dfrac{n!}{r!(n-r)!}$

This means n **factorial**. $n! = n \times (n-1) \times \ldots \times 3 \times 2 \times 1$

The expansion is valid as long as n is a **positive integer**.

$\binom{n}{r}$ or ${}^nC_r$ means n CHOOSE r – use the nCr function on your calculator to work it out.

Worked example

Find the first 3 terms, in ascending powers of x, of the binomial expansion of $(2 - 3x)^6$. Give each term in its simplest form. **(4 marks)**

$a = 2, b = -3x$

$(a + b)^6 = a^6 + \binom{6}{1}a^5b + \binom{6}{2}a^4b^2 + \ldots$

$(2 - 3x)^6 = 2^6 + \binom{6}{1} \times 2^5 \times (-3x) + \binom{6}{2} \times 2^4 \times (-3x)^2 + \ldots$

$= 64 - 576x + 2160x^2 + \ldots$

Be careful if either a or b is negative. Always use brackets if you are substituting anything more complicated than a positive whole number.

Problem solved!

$b = px$, so in the third term you need to square **all** of px.
$(px)^2 = p^2x^2$.
To work out $\binom{10}{2}$ or ${}^{10}C_2$ on your calculator,
type 10 nCr 2 =.
Make sure you **simplify** each term as much as possible. Don't leave any powers of numbers or multiplication signs in your final answer.

You will need to use problem-solving skills throughout your exam – **be prepared!**

Worked example

(a) Find the first 3 terms, in ascending powers of x, of the binomial expansion of $(1 + px)^{10}$, where p is a non-zero constant. Give each term in its simplest form. **(2 marks)**

$a = 1, b = px$

$(a + b)^{10} = a^{10} + \binom{10}{1}a^9b + \binom{10}{2}a^8b^2 + \ldots$

$(1 + px)^{10} = 1^{10} + \binom{10}{1} \times 1^9 \times px + \binom{10}{2} \times 1^8 \times (px)^2 + \ldots$

$= 1 + 10px + 45p^2x^2 + \ldots$

(b) Given that the coefficient of x^2 is 9 times the coefficient of x, find the value of p. **(2 marks)**

$45p^2 = 9(10p)$ so $45p^2 - 90p = 0$
$45p(p - 2) = 0$
$p = 0 \quad p = 2$

Now try this

1 Find the first 4 terms, in ascending powers of x, of each of these binomial expansions, giving each term in its simplest form.
(a) $(1 + 3x)^9$ **(3 marks)**
(b) $(2 + 5x)^4$ **(4 marks)**
(c) $(3 - x)^{12}$ **(4 marks)**

2 (a) Find the first 3 terms, in ascending powers of x, of the binomial expansion of $(2 + kx)^5$, where k is a constant. **(4 marks)**
(b) Given that the coefficient of x is 48, find the value of k. **(2 marks)**
(c) Write down the coefficient of x^2 **(1 mark)**

Solving binomial problems

You can use the binomial expansion to make **approximations** or to solve harder problems.

Worked example

Find the coefficient of x^5 in the binomial expansion of $\left(6 - \frac{x}{3}\right)^9$ **(2 marks)**

$a = 6$, $b = -\frac{x}{3}$, $n = 9$ so x^r term $= \binom{n}{r}a^{n-r}b^r$

x^5 term $= \binom{9}{5} \times 6^4 \times \left(-\frac{x}{3}\right)^5$

$= 126 \times 1296 \times \left(-\frac{1}{243}\right)x^5$

$= -672x^5$

The coefficient of x^5 is -672.

You can use this part of the binomial expansion of $(a + b)^n$ in the formulae booklet to find the x^r term without finding every term up to it:

$$\ldots + \binom{n}{r}a^{n-r}b^r + \ldots$$

The x part of each term comes from $b = -\frac{x}{3}$, so use $r = 5$ to get the x^5 term. Be careful with the fraction part:

$$\left(-\frac{x}{3}\right)^5 = \left(-\frac{1}{3}\right)^5 x^5$$

Binomial approximations

You can use a binomial expansion to **estimate** values. This is especially useful if x is **small**. If x is less than 1, then **larger power** of x get **smaller**. By ignoring large powers of x you can find a simple approximation. For example:

$(1 + x)^{100} \approx 1 + 100x + 4950x^2$

This means 'is approximately equal to'.

Write down the value of x you need to substitute. You can check your answer with a calculator. $1.05^6 = 1.340\,095\ldots$ ✓

Worked example

The first 4 terms of the binomial expansion of $\left(1 + \frac{x}{2}\right)^6$ are given below.

$\left(1 + \frac{x}{2}\right)^6 = 1 + 3x + \frac{15}{4}x^2 + \frac{5}{2}x^3 + \ldots$

Use the expansion to estimate the value of $(1.05)^6$ **(3 marks)**

If $x = 0.1$, then $\frac{x}{2} = 0.05$ and $\left(1 + \frac{x}{2}\right)^6 = (1.05)^6$

$(1.05)^6 \approx 1 + 3 \times (0.1) + \frac{15}{4} \times (0.1)^2 + \frac{5}{2} \times (0.1)^3 + \ldots$

$= 1 + 0.3 + 0.0375 + 0.0025 + \ldots$

$= 1.34$

Worked example

The first 4 terms of the binomial expansion of $(2 - x)^7$ are given below.

$(2 - x)^7 = 128 - 448x + 672x^2 + 560x^3 + \ldots$

If x is small, so that x^2 and higher powers can be ignored, show that $(1 + x)(2 - x)^7 \approx 128 - 320x$ **(2 marks)**

$(1 + x)(128 - 448x + \ldots)$

$= 128 + 128x - 448x - 448x^2 + \ldots$

$= 128 - 320x$ ignoring x^2 and higher powers

You are going to **ignore** x^2 and higher powers, so you only need to consider the first two terms of the expansion of $(2 - x)^7$. All the other terms would give you x^2 or higher terms when the brackets are multiplied out.

Now try this

1 In the binomial expansion of $(1 + 2x)^{30}$, the coefficients of x^3 and x^4 are p and q respectively.

(a) Show that $p = 32\,480$ **(1 mark)**

(b) Find the value of $\left(\frac{q}{p}\right)$ **(2 marks)**

2 (a) Find the first 4 terms, in ascending powers of x, of the binomial expansion of $\left(1 + \frac{x}{4}\right)^{12}$ **(4 marks)**

(b) Use your expansion to estimate the value of $(1.025)^{12}$, giving your answer to 4 decimal places. **(3 marks)**

Proof

If you need to **prove** a statement in your exam, you need to construct a **logical argument** to show that it is always true. You need to know these two methods of proof.

1 **Proof by deduction**
Also called **direct proof**. You use known facts and follow logical steps to reach a conclusion.

2 **Proof by exhaustion**
You consider **each of the possible cases** separately, in order to show that something is true in every case.

Worked example

x and y are rational numbers.
Prove that the mean of x and y is also a rational number. **(3 marks)**

Let $x = \dfrac{a}{b}$ and $y = \dfrac{c}{d}$, where a, b, c and d are integers. The mean of x and y is:

$$\frac{x + y}{2} = \frac{1}{2}\left(\frac{a}{b} + \frac{c}{d}\right) = \frac{1}{2}\left(\frac{ad}{bd} + \frac{bc}{bd}\right) = \frac{ad + bc}{2bd}$$

Since a, b, c and d are integers, $(ad + bc)$ and $2bd$ must also be integers, so $\dfrac{x + y}{2}$ is a rational number.

Proof checklist

✓ Write down any rules, information or assumptions you need to use.

✓ Show each step of your working clearly.

✓ Make sure your steps follow logically from each other.

✓ Write down what you have proved at the end of your working.

← Rational numbers are numbers that can be written in the form $\dfrac{p}{q}$, where p and q are integers.

Problem solved!

You can prove this result by considering the cases where x and y are positive or negative separately. You know you have covered **all possible cases** because at least one of the following statements must be true:
- x is negative
- y is negative
- both x and y are non-negative.

Remember to write down what you have proved at the end of your working.

You will need to use problem-solving skills throughout your exam – **be prepared!**

Worked example

Prove that $(xy + 1)^2 + (x - 1)^2 + (y - 1)^2 \geqslant 1$ for all real values of x and y. **(4 marks)**

Each of $(xy + 1)^2$, $(x - 1)^2$ and $(y - 1)^2$ is $\geqslant 0$ for all real values of x and y.

If $x < 0$ then $(x - 1)^2 > 1$

If $y < 0$ then $(y - 1)^2 > 1$

If x and y are both $\geqslant 0$ then $(xy + 1)^2 \geqslant 1$

All three terms are $\geqslant 0$ and at least one must be $\geqslant 1$, so the whole expression must be $\geqslant 1$, as required.

Now try this

1 $f(x) = 2x^3 + 5x^2 - 4$
A student is attempting to use the factor theorem to prove that $(x + 2)$ is a factor of $f(x)$. The student writes the following working:

> $f(-2) = 2(-2)^3 + 5(-2)^2 - 4$
> $\qquad = -16 + 20 - 4$
> $\qquad = 0$

(a) Explain why this proof is incomplete. **(1 mark)**
(b) Write an additional line of working to complete the proof. **(1 mark)**

2 (a) Prove by exhaustion that $2n^2 + 11$ is a prime number for all $n \in \mathbb{N}$, $n \leqslant 10$. **(2 marks)**

(b) By means of a counterexample disprove the statement:
$n^2 + n + 17$ is a prime number for all $n \in \mathbb{N}$. **(2 marks)**

3 Prove that $(4 - x)^2 \geqslant 7 - 2x$ for all real values of x. **(3 marks)**

Cosine rule

The cosine rule applies to **any triangle**. You usually use the cosine rule when you know **two sides** and the **angle between them (SAS)** or when you are given **three sides** and you want to work out an **angle (SSS)**.

1 $a^2 = b^2 + c^2 - 2bc\cos A$

Use this version to find a missing side.

2 $\cos A = \dfrac{b^2 + c^2 - a^2}{2bc}$

This version is useful for finding a missing angle.

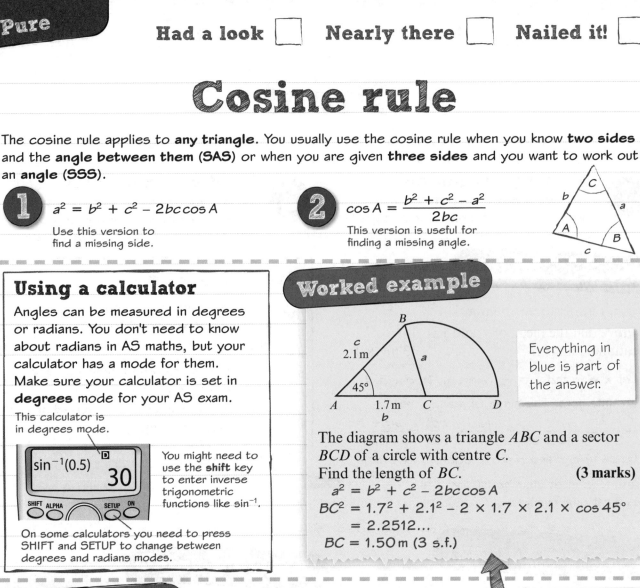

Using a calculator

Angles can be measured in degrees or radians. You don't need to know about radians in AS maths, but your calculator has a mode for them. Make sure your calculator is set in **degrees** mode for your AS exam.

This calculator is in degrees mode.

$\sin^{-1}(0.5)$ **30**

SHIFT ALPHA   SETUP ON

You might need to use the **shift** key to enter inverse trigonometric functions like $\sin^{-1}$.

On some calculators you need to press SHIFT and SETUP to change between degrees and radians modes.

Worked example

The diagram shows a triangle ABC and a sector BCD of a circle with centre C.
Find the length of BC.   **(3 marks)**

$a^2 = b^2 + c^2 - 2bc\cos A$

$BC^2 = 1.7^2 + 2.1^2 - 2 \times 1.7 \times 2.1 \times \cos 45°$

$= 2.2512...$

$BC = 1.50\,\text{m}$ (3 s.f.)

Everything in blue is part of the answer.

You know two sides and the angle between them (SAS) so you can use the cosine rule to find the opposite side. Label the sides with the lower case letter of the **opposite angle** and write out the formula before you substitute.

Worked example

In the triangle ABC, $AB = 15\,\text{cm}$, $BC = 9\,\text{cm}$ and $CA = 10\,\text{cm}$. Find the size of angle C, giving your answer to the nearest degree.   **(3 marks)**

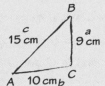

$\cos C = \dfrac{a^2 + b^2 - c^2}{2ab}$

$= \dfrac{9^2 + 10^2 - 15^2}{2 \times 9 \times 10} = -0.2444...$

$C = \cos^{-1}(-0.2444...) = 104°$

If no diagram is given in the question you can sketch one. You know three sides (SSS) so you can use the cosine rule to find any angle in the triangle. Be careful with the order. You add the squares of the sides **adjacent** to the angle, and subtract the square of the **opposite** side.

Now try this

The diagram shows two triangles PQR and PRS.
$\angle RSP = 50°$. Find
(a) the length of PR **(3 marks)**
(b) the size of $\angle PQR$, giving your answer in degrees to 1 decimal place.   **(3 marks)**

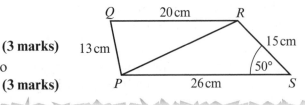

Sine rule

You need to **learn** the sine rule for your exam. It applies to **any triangle**. The sine rule is useful when you know **two angles**, or when you know a side and the **opposite** angle.

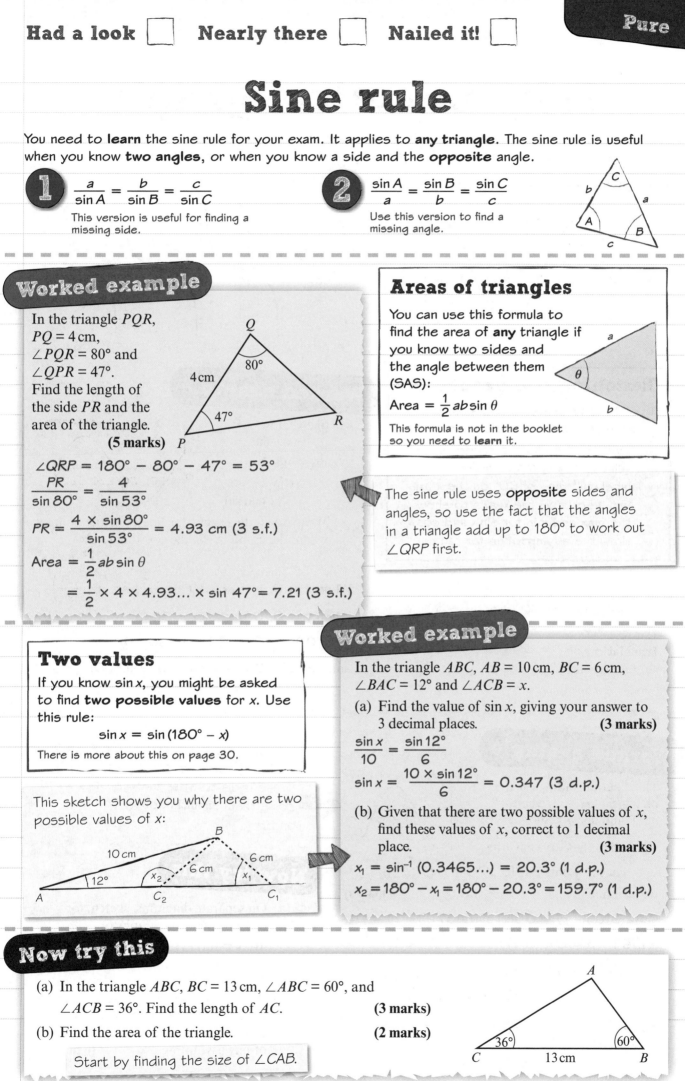

1　$\dfrac{a}{\sin A} = \dfrac{b}{\sin B} = \dfrac{c}{\sin C}$

This version is useful for finding a missing side.

2　$\dfrac{\sin A}{a} = \dfrac{\sin B}{b} = \dfrac{\sin C}{c}$

Use this version to find a missing angle.

Worked example

In the triangle PQR, $PQ = 4\,\text{cm}$, $\angle PQR = 80°$ and $\angle QPR = 47°$. Find the length of the side PR and the area of the triangle.

(5 marks)

$\angle QRP = 180° - 80° - 47° = 53°$

$\dfrac{PR}{\sin 80°} = \dfrac{4}{\sin 53°}$

$PR = \dfrac{4 \times \sin 80°}{\sin 53°} = 4.93\text{ cm (3 s.f.)}$

Area $= \dfrac{1}{2}ab\sin\theta$

$\quad\quad = \dfrac{1}{2} \times 4 \times 4.93... \times \sin 47° = 7.21\text{ (3 s.f.)}$

Areas of triangles

You can use this formula to find the area of **any** triangle if you know two sides and the angle between them (SAS):

Area $= \dfrac{1}{2}ab\sin\theta$

This formula is not in the booklet so you need to **learn** it.

The sine rule uses **opposite** sides and angles, so use the fact that the angles in a triangle add up to 180° to work out $\angle QRP$ first.

Two values

If you know $\sin x$, you might be asked to find **two possible values** for x. Use this rule:

$$\sin x = \sin(180° - x)$$

There is more about this on page 30.

This sketch shows you why there are two possible values of x:

Worked example

In the triangle ABC, $AB = 10\,\text{cm}$, $BC = 6\,\text{cm}$, $\angle BAC = 12°$ and $\angle ACB = x$.

(a) Find the value of $\sin x$, giving your answer to 3 decimal places. **(3 marks)**

$\dfrac{\sin x}{10} = \dfrac{\sin 12°}{6}$

$\sin x = \dfrac{10 \times \sin 12°}{6} = 0.347\text{ (3 d.p.)}$

(b) Given that there are two possible values of x, find these values of x, correct to 1 decimal place. **(3 marks)**

$x_1 = \sin^{-1}(0.3465...) = 20.3°\text{ (1 d.p.)}$

$x_2 = 180° - x_1 = 180° - 20.3° = 159.7°\text{ (1 d.p.)}$

Now try this

(a) In the triangle ABC, $BC = 13\,\text{cm}$, $\angle ABC = 60°$, and $\angle ACB = 36°$. Find the length of AC. **(3 marks)**

(b) Find the area of the triangle. **(2 marks)**

Start by finding the size of $\angle CAB$.

28

Trigonometric graphs

You need to be able to sketch the graphs of **sin**, **cos** and **tan**, and **transformations** of them. If you want to recap transformations of graphs, have a look at pages 13 and 14.

$y = \sin x$ and $y = \cos x$

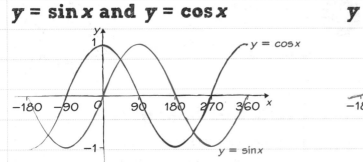

$y = \tan x$

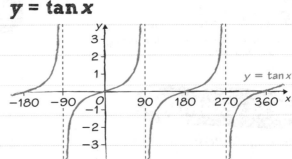

Sketching trig graphs

If you have to sketch a trigonometric graph in your exam, make sure you:

☑ pay attention to the **range** of values for x in the question

☑ label multiples of 90° on the x-axis

☑ put a scale on the y-axis to show the **max** and **min** for $\sin x$ and $\cos x$

☑ draw the **asymptotes** for $\tan x$.

You can write cos 30° exactly as a surd.

The graph of $y = (\cos x + 30°)$ is a **translation** of $y = \cos x$. The graph moves 30° to the **left**.

Worked example

(a) Sketch, for $0 \le x \le 360°$, the graph of $y = \cos(x + 30°)$
(2 marks)

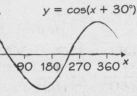

(b) Write down the exact coordinates of the points where the graph meets the coordinate axes.
(3 marks)

When $x = 0$, $y = \cos 30° = \dfrac{\sqrt{3}}{2}$, so $\left(0, \dfrac{\sqrt{3}}{2}\right)$

When $y = 0$, $0 = \cos(x + 30°)$:

$90° - 30° = 60°$, so $(60°, 0)$

and $270° - 30° = 240°$, so $(240°, 0)$

Worked example

The diagram shows a sketch of $y = \sin(ax - b)$, where $a > 0$ and $0 < b < 360°$.

Given that the curve cuts the x-axis at the points $P\,(216°, 0)$ and $Q\,(576°, 0)$, find a and b. **(4 marks)**

$\sin(a(216°) - b) = 0$ and $\sin(a(576°) - b) = 0$

$a(216°) - b = 0$ ①

$a(576°) - b = 180°$ ②

② − ①: $(360°)a = 180°$ so $a = \frac{1}{2}$

Substituting into ①: $\frac{1}{2}(216°) - b = 0$
so $b = 108°$

If $\sin(ax - b) = 0$, then $ax - b = 0$, or 180°, or 360° and so on. You can use these facts to write two equations and solve them **simultaneously** to find a and b.

Now try this

(a) On separate diagrams, sketch, for $0 \le x \le 360°$, the graphs of

   (i) $y = \sin(2x)$

   (ii) $y = \tan(x + 90°)$ **(4 marks)**

(b) Write down the coordinates of any points where the curves meet the coordinate axes and the equations of any asymptotes. **(6 marks)**

Trigonometric equations 1

You can solve an equation involving **sin**, **cos** or **tan**. You need to be really careful because these equations can have **multiple solutions**. You will be given a **range** (or **interval**) of values for x. You need to find values of x that are in that range.

Using graphs to find solutions

This graph shows the solutions to the equation $\cos x = -\frac{1}{2}$ in the range $-180° \leq x \leq 360°$.

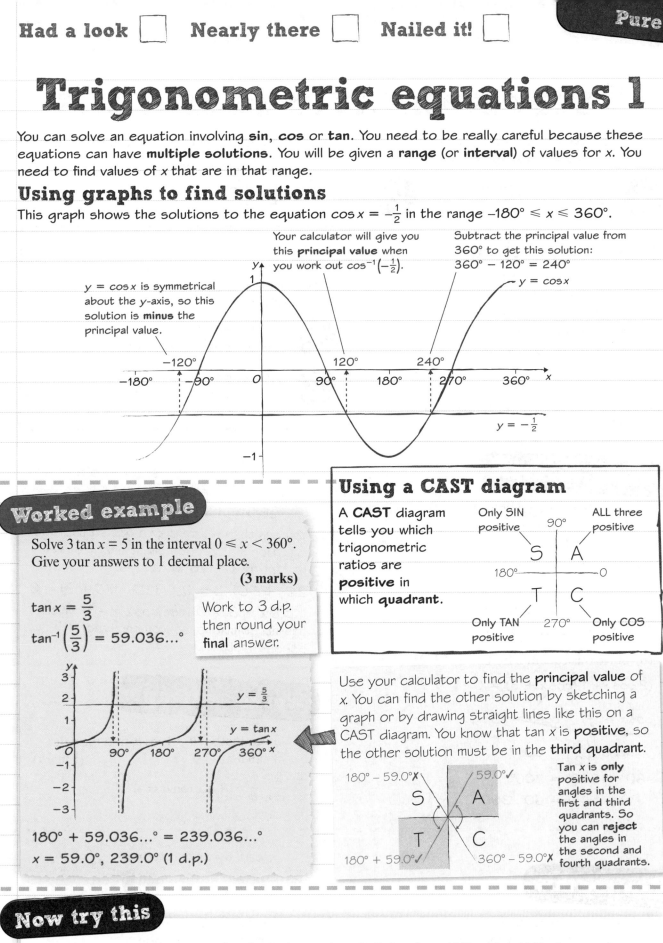

Your calculator will give you this **principal value** when you work out $\cos^{-1}\left(-\frac{1}{2}\right)$.

Subtract the principal value from 360° to get this solution: $360° - 120° = 240°$

$y = \cos x$ is symmetrical about the y-axis, so this solution is **minus** the principal value.

$y = \cos x$

$y = -\frac{1}{2}$

Worked example

Solve $3\tan x = 5$ in the interval $0 \leq x < 360°$. Give your answers to 1 decimal place. **(3 marks)**

$\tan x = \dfrac{5}{3}$

$\tan^{-1}\left(\dfrac{5}{3}\right) = 59.036...°$

Work to 3 d.p. then round your **final** answer.

$y = \frac{5}{3}$

$y = \tan x$

$180° + 59.036...° = 239.036...°$

$x = 59.0°, 239.0°$ (1 d.p.)

Using a CAST diagram

A **CAST** diagram tells you which trigonometric ratios are **positive** in which **quadrant**.

Only SIN positive | ALL three positive
S | A
180° ——— 0
T | C
Only TAN positive | Only COS positive
270°
90°

Use your calculator to find the **principal value** of x. You can find the other solution by sketching a graph or by drawing straight lines like this on a CAST diagram. You know that $\tan x$ is **positive**, so the other solution must be in the **third quadrant**.

$180° - 59.0°$ ✗ | $59.0°$ ✓
S | A
T | C
$180° + 59.0°$ ✓ | $360° - 59.0°$ ✗

Tan x is **only** positive for angles in the first and third quadrants. So you can **reject** the angles in the second and fourth quadrants.

Now try this

1 (a) Sketch the graph of $y = \sin x$ in the interval $0 \leq x < 360°$. **(2 marks)**

   (b) Find the values of x in the interval $0 \leq x \leq 360°$ for which $\sin x = -0.3$. Give your answers correct to 1 decimal place. **(3 marks)**

2 Solve, for $-180° \leq \theta < 180°$, the equation

   (a) $3\cos\theta = 1$ **(3 marks)**

   (b) $\tan\theta + 2 = 0$ **(3 marks)**

There will be **two solutions** to each equation. Find one using your calculator, then sketch the graph to find the other.

Trigonometric identities

You might need to use one of these two trigonometric identities to **simplify** a trig equation before solving it. They are true for **all values** of x or θ.

— $\sin^2 \theta$ means $(\sin\theta)^2$.

1 $\tan\theta \equiv \dfrac{\sin\theta}{\cos\theta}$ **2** $\sin^2\theta + \cos^2\theta \equiv 1$

Quadratic equations

If an equation involves $\sin^2\theta$ and $\sin\theta$ (or $\cos^2\theta$ and $\cos\theta$) then it is a quadratic. You can solve it by factorising:

$2\sin^2\theta - 3\sin\theta - 2 = 0$

$(2\sin\theta + 1)(\sin\theta - 2) = 0$

$\sin\theta = -\frac{1}{2}$ $\sin\theta = 2$

No solutions exist to $\sin\theta = 2$, so you would only solve $\sin\theta = -\frac{1}{2}$.

Golden rules

When finding solutions to **quadratic trigonometric** equations, remember these golden rules:

1 Write everything in terms of $\sin^2\theta$ and $\sin\theta$ (or $\cos^2\theta$ and $\cos\theta$).

2 Solutions to $\sin x = k$ and $\cos x = k$ **only exist** if $-1 \leqslant k \leqslant 1$. Solutions to $\tan x = k$ exist for **any value** of k.

Worked example

(a) Show that the equation $5\cos x = 1 + 2\sin^2 x$ can be written in the form
$2\cos^2 x + 5\cos x - 3 = 0$ **(2 marks)**

$5\cos x = 1 + 2\sin^2 x$

$\qquad = 1 + 2(1 - \cos^2 x)$

$\qquad = 1 + 2 - 2\cos^2 x$

$2\cos^2 x + 5\cos x - 3 = 0$

You can use $\sin^2 x + \cos^2 x = 1$ to rewrite $\sin^2 x$ in terms of $\cos^2 x$:

$\sin^2 x = 1 - \cos^2 x$

You can then rearrange to get a quadratic equation in $\cos x$. You might find it easier to factorise if you write it as:

$2C^2 + 5C - 3 = 0 \rightarrow (2C - 1)(C + 3) = 0$

If $\cos x + 3 = 0$ then $\cos x = -3$, which has **no solutions**, so you can ignore the second factor.

(b) Solve this equation for $0 \leqslant x < 360°$. **(3 marks)**

$(2\cos x - 1)(\cos x + 3) = 0$

$\cos x = \dfrac{1}{2}$ ~~$\cos x = -3$~~

$\cos^{-1}\left(\dfrac{1}{2}\right) = 60°$

$-60° + 360° = 300°$

$\qquad x = 60°, 300°$

Worked example

Given that $\sin\theta = 4\cos\theta$, find the value of $\tan\theta$.   **(1 mark)**

$\dfrac{\sin\theta}{\cos\theta} = 4$ so $\tan\theta = 4$

Start by writing $\tan x$ as $\dfrac{\sin x}{\cos x}$

Now try this

1 Find all the solutions, in the interval $0 \leqslant x < 360°$, of the equation
$$3\cos^2 x - 9 = 11\sin x$$
giving each solution correct to 1 decimal place.   **(6 marks)**

Use $\cos^2 x = 1 - \sin^2 x$ to get a quadratic equation in $\sin x$.

2 (a) Show that the equation $5\sin x = 2\tan x$ can be written in the form
$$\sin x(5\cos x - 2) = 0$$ **(2 marks)**

(b) Solve, for $0 \leqslant x < 360°$, $5\sin x = 2\tan x$ **(4 marks)**

Either $\sin x = 0$, or $5\cos x - 2 = 0$. Both these factors will give you solutions.

Trigonometric equations 2

You need to be careful if a trigonometric equation involves a **function** of x or θ.

Worked example

Find the exact solutions of the equation
$$\cos(\theta - 50°) = 0.5$$
in the interval $0 \leqslant \theta < 180°$. **(3 marks)**

$0 \leqslant \theta < 180°$ so $-50° \leqslant \theta - 50° < 130°$

Let $Z = \theta - 50°$

$\cos Z = 0.5$, $-50° \leqslant Z < 130°$

So $Z = \cos^{-1}(0.5) = 60°$ in range ✓

or $Z = -60°$ not in range ✗

or $Z = 360° - 60° = 300°$ not in range ✗

So $\theta - 50° = 60°$

So $\theta = 110°$

Transforming the range

The safest way to solve an equation involving sin, cos or tan of $(x + b)$ or (ax) is to **transform the range**.

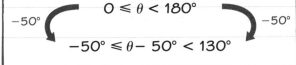

$$0 \leqslant \theta < 180°$$
$$-50° \qquad -50°$$
$$-50° \leqslant \theta - 50° < 130°$$

If $Z = \theta - 50°$, you need to find all the values of Z such that $\cos Z = 0.5$ in the range $-50° \leqslant Z < 130°$. You can then find the corresponding value of θ for each solution.

Worked example

The graph of $y = \sin(3x)$ is a horizontal stretch of the graph of $y = \sin x$ with scale factor $\frac{1}{3}$.
This sketch shows the solutions of the equation.

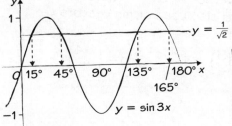

Make sure you remember to transform your solutions for $3x$ back into solutions for x at the end, and double check that they all lie within $0 \leqslant x < 180°$.

Solve, for $0 \leqslant x < 180°$, the equation $\sin(3x) = \dfrac{1}{\sqrt{2}}$

(6 marks)

$0 \leqslant x < 180°$ so $0 \leqslant 3x < 540°$

Let $Z = 3x$

$\sin Z = \dfrac{1}{\sqrt{2}}$, $0 \leqslant Z < 540°$

So $Z = \sin^{-1}\left(\dfrac{1}{\sqrt{2}}\right) = 45°$ in range ✓

or $Z = 45° + 360° = 405°$ in range ✓

or $Z = 45° + 720° = 765°$ not in range ✗

or $Z = 180° - 45° = 135°$ in range ✓

or $Z = 180° - 45° + 360° = 495°$ in range ✓

or $Z = 180° - 45° + 720° = 855°$ not in range ✗

So $3x = 45°$ or $405°$ or $135°$ or $495°$

So $x = 15°$ or $135°$ or $45°$ or $165°$

Now try this

1 Solve, for $0 \leqslant x < 360°$

(a) $\sin(x - 40°) = -\frac{1}{2}$ **(4 marks)**

(b) $\cos(2x) = \dfrac{\sqrt{3}}{2}$ **(4 marks)**

2 Find all the solutions of the equation $\cos^2(x + 30°) = \frac{1}{4}$ in the range $-180° \leqslant x \leqslant 180°$. **(6 marks)**

You need to consider the positive and negative square roots separately. This is like solving two separate equations:
$\cos(x + 30°) = \frac{1}{2}$ and
$\cos(x + 30°) = -\frac{1}{2}$

32

Vectors

Vectors can be described using column vectors, or using **i, j** notation:

$$\overrightarrow{XY} = \begin{pmatrix} 3 \\ -1 \end{pmatrix} = 3\mathbf{i} - \mathbf{j}$$

i and **j** are **perpendicular unit vectors**.

$$\mathbf{j} = \begin{pmatrix} 0 \\ 1 \end{pmatrix} \qquad \mathbf{i} = \begin{pmatrix} 1 \\ 0 \end{pmatrix}$$

O

Position or direction?

It is useful to distinguish between **position vectors** and **direction vectors**.

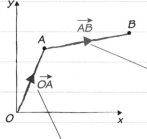

The direction vector $\overrightarrow{AB}$ tells you the direction and distance from A to B.

A position vector starts at the **origin**. $\overrightarrow{OA}$ tells you the position of point A.

Magnitude

You can find the magnitude of a vector using Pythagoras' theorem.

$$\left| \overrightarrow{AB} \right| = \left| \begin{pmatrix} 2 \\ -4 \end{pmatrix} \right| = |2\mathbf{i} - 4\mathbf{j}|$$

$$= \sqrt{2^2 + 4^2} = 2\sqrt{5}$$

Ignore minus signs when calculating the magnitude of a vector.

✓ **unit vectors** have magnitude 1.

✓ The **distance** between two points A and B is the magnitude of the vector $\overrightarrow{AB}$.

Worked example

The points *P* and *Q* have position vectors $3\mathbf{i} + 4\mathbf{j}$ and $-\mathbf{i} + 5\mathbf{j}$ respectively.

(a) Find the vector $\overrightarrow{PQ}$.   **(2 marks)**

$\overrightarrow{PQ} = \overrightarrow{OQ} - \overrightarrow{OP}$

$\quad = (-1 - 3)\mathbf{i} + (5 - 4)\mathbf{j}$

$\quad = -4\mathbf{i} + \mathbf{j}$

(b) Find the distance *PQ*.   **(1 mark)**

$\left| \overrightarrow{PQ} \right| = \sqrt{-4^2 + 1^2}$

$\quad = \sqrt{17}$

(c) Find a unit vector in the direction of $\overrightarrow{PQ}$.
   (1 mark)

$\dfrac{1}{\sqrt{17}} \overrightarrow{PQ} = \dfrac{1}{\sqrt{17}}(-4\mathbf{i} + \mathbf{j})$

$\quad = -\dfrac{4}{\sqrt{17}}\mathbf{i} + \dfrac{1}{\sqrt{17}}\mathbf{j}$

$$\overrightarrow{PQ} = \begin{pmatrix} \text{Position} \\ \text{vector of } Q \end{pmatrix} - \begin{pmatrix} \text{Position} \\ \text{vector of } P \end{pmatrix}$$

You could also use column vectors to subtract:

$$\begin{pmatrix} -1 \\ 5 \end{pmatrix} - \begin{pmatrix} 3 \\ 4 \end{pmatrix} = \begin{pmatrix} -1 - 3 \\ 5 - 4 \end{pmatrix} = \begin{pmatrix} -4 \\ 1 \end{pmatrix}$$

You can't write in bold in your exam! You can underline vectors to make them clearer. If you're writing the vector between two points, you should draw an arrow over the top. $\overrightarrow{PQ}$ is the **direction vector** from *P* to *Q*, whereas *PQ* is the **line segment** between *P* and *Q*.

Now try this

1 The points *A* and *B* have position vectors $4\mathbf{i} - 2\mathbf{j}$ and $5\mathbf{i} + 2\mathbf{j}$ respectively.

   (a) Find the vector $\overrightarrow{AB}$.   **(2 marks)**

   (b) Write down the vector $\overrightarrow{BA}$.   **(1 mark)**

2 Find a unit vector in the direction of $\begin{pmatrix} 2 \\ -10 \end{pmatrix}$
   (2 marks)

Solving vector problems

You can use these formulae to find areas of **triangles** and **parallelograms** in vector questions:

1   Area of triangle ABC = $\frac{1}{2}|\mathbf{a}||\mathbf{b}|\sin\theta$

2   Area of parallelogram ABCD = $|\mathbf{a}||\mathbf{b}|\sin\theta$

The area of the parallelogram is **twice** the area of the triangle.

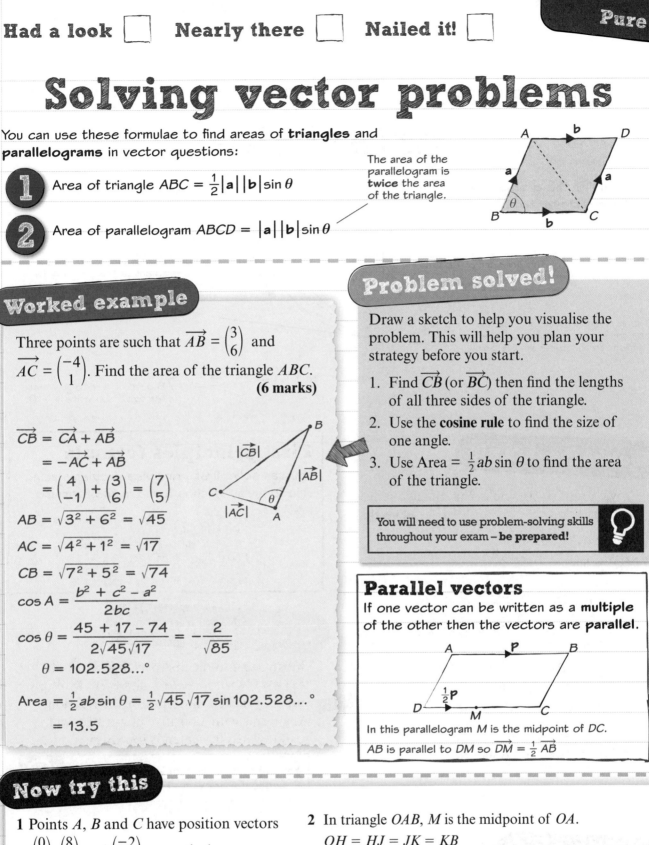

Worked example

Three points are such that $\overrightarrow{AB} = \binom{3}{6}$ and $\overrightarrow{AC} = \binom{-4}{1}$. Find the area of the triangle ABC.

(6 marks)

$\overrightarrow{CB} = \overrightarrow{CA} + \overrightarrow{AB}$

$\quad = -\overrightarrow{AC} + \overrightarrow{AB}$

$\quad = \binom{4}{-1} + \binom{3}{6} = \binom{7}{5}$

$AB = \sqrt{3^2 + 6^2} = \sqrt{45}$

$AC = \sqrt{4^2 + 1^2} = \sqrt{17}$

$CB = \sqrt{7^2 + 5^2} = \sqrt{74}$

$\cos A = \dfrac{b^2 + c^2 - a^2}{2bc}$

$\cos\theta = \dfrac{45 + 17 - 74}{2\sqrt{45}\sqrt{17}} = -\dfrac{2}{\sqrt{85}}$

$\theta = 102.528...°$

Area $= \frac{1}{2}ab\sin\theta = \frac{1}{2}\sqrt{45}\sqrt{17}\sin 102.528...°$

$\quad = 13.5$

Problem solved!

Draw a sketch to help you visualise the problem. This will help you plan your strategy before you start.

1. Find $\overrightarrow{CB}$ (or $\overrightarrow{BC}$) then find the lengths of all three sides of the triangle.
2. Use the **cosine rule** to find the size of one angle.
3. Use Area = $\frac{1}{2}ab\sin\theta$ to find the area of the triangle.

You will need to use problem-solving skills throughout your exam – **be prepared!**

Parallel vectors

If one vector can be written as a **multiple** of the other then the vectors are **parallel**.

In this parallelogram M is the midpoint of DC.

AB is parallel to DM so $\overrightarrow{DM} = \frac{1}{2}\overrightarrow{AB}$

Now try this

1 Points A, B and C have position vectors $\binom{0}{4}$, $\binom{8}{3}$ and $\binom{-2}{5}$ respectively.

(a) Find the vectors $\overrightarrow{AB}$, $\overrightarrow{BC}$ and $\overrightarrow{CA}$. **(3 marks)**

(b) Find the area of triangle ABC. **(4 marks)**

Point D is such that the points A, B, C and D form the vertices of a parallelogram.

(c) Write down the area of the parallelogram. **(1 mark)**

(d) Find the position vector of three possible positions of D. **(4 marks)**

2 In triangle OAB, M is the midpoint of OA.

$OH = HJ = JK = KB$

S and T divide the line AB into three equal segments.

$\overrightarrow{OM} = \mathbf{a}$ and $\overrightarrow{OH} = \mathbf{b}$

(a) Prove that $\overrightarrow{HA}$, $\overrightarrow{JS}$ and $\overrightarrow{KT}$ are all parallel. **(6 marks)**

(b) State the ratio $KT : JS : HA$ **(1 mark)**

Differentiating from first principles

You can find the **gradient** of a curve at a point A by considering the gradient of the chord AB as B gets closer to A.

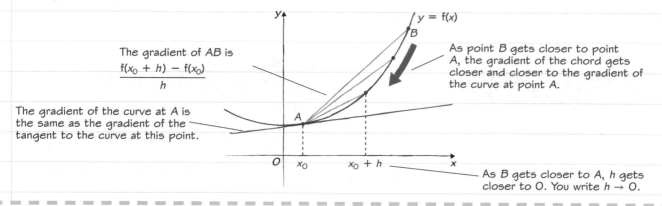

The gradient of AB is
$$\frac{f(x_0 + h) - f(x_0)}{h}$$

The gradient of the curve at A is the same as the gradient of the tangent to the curve at this point.

As point B gets closer to point A, the gradient of the chord gets closer and closer to the gradient of the curve at point A.

As B gets closer to A, h gets closer to O. You write $h \rightarrow 0$.

Worked example

Prove from first principles that the derivative of $4x^3$ is $12x^2$. **(4 marks)**

$f(x) = 4x^3$

$f'(x) = \lim_{h \to 0} \dfrac{f(x + h) - f(x)}{h}$

$= \lim_{h \to 0} \dfrac{4(x + h)^3 - 4x^3}{h}$

$= \lim_{h \to 0} \dfrac{4(x^3 + 3x^2h + 3xh^2 + h^3) - 4x^3}{h}$

$= \lim_{h \to 0} \dfrac{12x^2h + 12xh^2 + 4h^3}{h}$

$= \lim_{h \to 0} (12x^2 + 12xh + 4h^2)$

As $h \rightarrow 0$, then $12xh \rightarrow 0$ and $4h^2 \rightarrow 0$.

So $f'(x) = 12x^2$, as required.

First principles formula

You can solve first principles problems using this formula, which is found in the formulae booklet.

First Principles

$f'(x) = \lim_{h \to 0} \dfrac{f(x + h) - f(x)}{h}$

Problem solved!

Substitute into the first principles formula. If $f(x) = 4x^3$ then $f(x + h) = 4(x + h)^3$. Multiply out the brackets and simplify the fraction. Any terms with a positive power of h will tend towards 0 as h tends towards 0.

You will need to use problem-solving skills throughout your exam – **be prepared!**

Now try this

1 Given that $f(x) = x^2$, use differentiation from first principles to find $f'(7)$. **(5 marks)**

The question states "first principles" so you need to use the formula, or write an expression for the chord AB. Make sure you write out which terms in the expression $\rightarrow 0$ as $h \rightarrow 0$.

2 The points A and B with x-coordinates 5 and $5 + h$ respectively lie on the curve with equation $y = 2x^2 + 3x$.

(a) Show that the gradient of AB is $23 + 2h$. **(3 marks)**

(b) Deduce the gradient of the tangent to the curve at A. **(1 mark)**

Differentiation 1

You can **differentiate** a function to find its **derivative** or **gradient function**.

The derivative is written as f'(x) or $\dfrac{dy}{dx}$

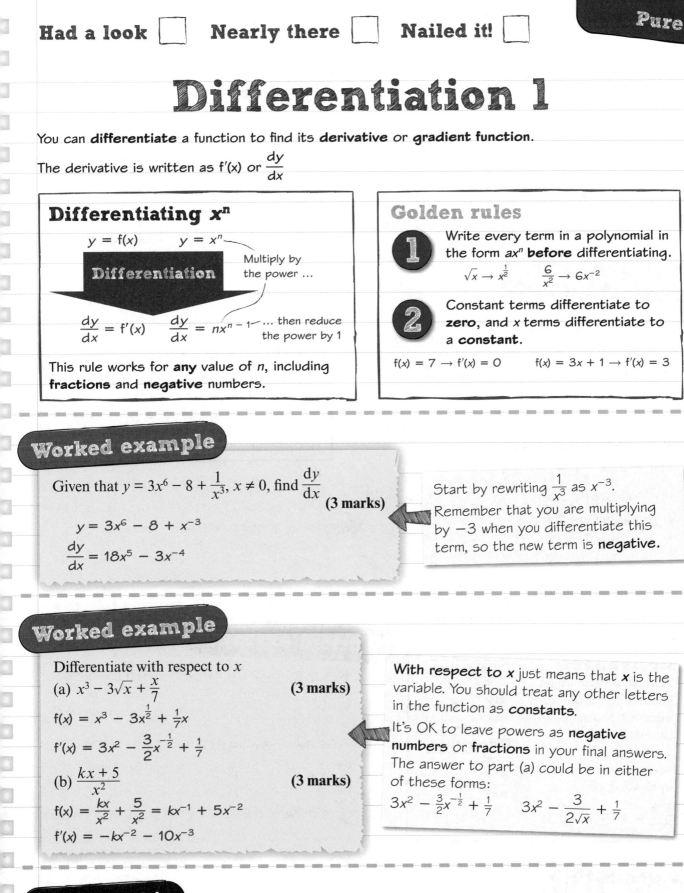

Differentiating x^n

$y = f(x)$ $y = x^n$

Differentiation

Multiply by the power ...

$\dfrac{dy}{dx} = f'(x)$ $\dfrac{dy}{dx} = nx^{n-1}$ ... then reduce the power by 1

This rule works for **any** value of n, including **fractions** and **negative** numbers.

Golden rules

1 Write every term in a polynomial in the form ax^n **before** differentiating.

$\sqrt{x} \to x^{\frac{1}{2}}$ $\dfrac{6}{x^2} \to 6x^{-2}$

2 Constant terms differentiate to **zero**, and x terms differentiate to a **constant**.

$f(x) = 7 \to f'(x) = 0$ $f(x) = 3x + 1 \to f'(x) = 3$

Worked example

Given that $y = 3x^6 - 8 + \dfrac{1}{x^3}$, $x \neq 0$, find $\dfrac{dy}{dx}$ **(3 marks)**

$y = 3x^6 - 8 + x^{-3}$

$\dfrac{dy}{dx} = 18x^5 - 3x^{-4}$

Start by rewriting $\dfrac{1}{x^3}$ as x^{-3}.
Remember that you are multiplying by -3 when you differentiate this term, so the new term is **negative**.

Worked example

Differentiate with respect to x

(a) $x^3 - 3\sqrt{x} + \dfrac{x}{7}$ **(3 marks)**

$f(x) = x^3 - 3x^{\frac{1}{2}} + \frac{1}{7}x$

$f'(x) = 3x^2 - \frac{3}{2}x^{-\frac{1}{2}} + \frac{1}{7}$

(b) $\dfrac{kx + 5}{x^2}$ **(3 marks)**

$f(x) = \dfrac{kx}{x^2} + \dfrac{5}{x^2} = kx^{-1} + 5x^{-2}$

$f'(x) = -kx^{-2} - 10x^{-3}$

With respect to x just means that x is the variable. You should treat any other letters in the function as **constants**.

It's OK to leave powers as **negative numbers** or **fractions** in your final answers. The answer to part (a) could be in either of these forms:

$3x^2 - \frac{3}{2}x^{-\frac{1}{2}} + \frac{1}{7}$ $3x^2 - \dfrac{3}{2\sqrt{x}} + \frac{1}{7}$

Now try this

1 Given that $y = \dfrac{(x + 3)^2}{x}$, $x \neq 0$, find $\dfrac{dy}{dx}$ **(4 marks)**

Multiply out the brackets, then write the function in the form $y = ax + b + cx^{-1}$ before differentiating.

2 (a) Write $\dfrac{2 + 5\sqrt{x}}{x}$ in the form $2x^p + 5x^q$, where p and q are constants.   **(2 marks)**

(b) Given that $y = 3x^2 + 1 - \dfrac{2 + 5\sqrt{x}}{x}$, find $\dfrac{dy}{dx}$ **(4 marks)**

Differentiation 2

You can use the **derivative** or **gradient function** to find the **rate of change** of a function, or the gradient of a curve.

This curve has equation $y = x^3 + 5x^2$. Its gradient function has equation $\frac{dy}{dx} = 3x^2 + 10x$. You can find the **gradient** at any point on the graph by substituting the x-coordinate at that point into the gradient function.

At the point P:

$x = 2$

$\frac{dy}{dx} = 3(2)^2 + 10(2) = 12 + 20 = 32$

Gradient at $P = 32$

Worked example

$f(x) = (10 + 2\sqrt{x})^2, x > 0$

(a) Show that $f(x) = 100 + k\sqrt{x} + 4x$, where k is a constant to be found. **(2 marks)**

$f(x) = 10^2 + 20\sqrt{x} + 20\sqrt{x} + (2\sqrt{x})^2$

$\quad = 100 + 40\sqrt{x} + 4x \qquad k = 40$

(b) Find $f'(x)$. **(2 marks)**

$f(x) = 100 + 40x^{\frac{1}{2}} + 4x$

$f'(x) = 20x^{-\frac{1}{2}} + 4$

(c) Evaluate $f'(25)$. **(1 mark)**

$f'(25) = 20\left(25^{-\frac{1}{2}}\right) + 4$

$\quad = 20\left(\frac{1}{5}\right) + 4$

$\quad = 4 + 4 = 8$

Evaluating $f'(x)$

$f'(x)$ tells you the **rate of change** of the function for a given value of x.

You can calculate $f'(x)$ for a given value of x by substituting that value of x into the derivative.

$25^{-\frac{1}{2}} = \frac{1}{25^{\frac{1}{2}}} = \frac{1}{\sqrt{25}} = \frac{1}{5}$

For a reminder about using the index laws to simplify powers have a look at page 1.

Second-order derivatives

You can differentiate **twice** to find the **second-order** derivative.

You write the second-order derivative as $\frac{d^2y}{dx^2}$ or $f''(x)$.

$y = 5x^3$

│ Differentiate │

$\frac{dy}{dx} = 15x^2$

│ Differentiate │

$\frac{d^2y}{dx^2} = 30x$

Worked example

Given that $y = 8\sqrt{x} - 3x^2 + 5x, x > 0$,

find $\frac{d^2y}{dx^2}$ **(4 marks)**

$y = 8x^{\frac{1}{2}} - 3x^2 + 5x$

$\frac{dy}{dx} = 4x^{-\frac{1}{2}} - 6x + 5$

$\frac{d^2y}{dx^2} = -2x^{-\frac{3}{2}} - 6$

Now try this

1 $f(x) = 3x^3 + 5x, x > 0$

   (a) Differentiate to find $f'(x)$. **(2 marks)**

   (b) Given that $f'(x) = 41$, find the value of x. **(3 marks)**

2 Given that $y = 2x^2 + 4x^{-2}$, find $\frac{d^2y}{dx^2}$ **(4 marks)**

3 The curve C has equation $y = x(x - 1)(x + 3)$

   (a) Find $\frac{dy}{dx}$ **(2 marks)**

   (b) Sketch C, showing each point where C meets the x-axis. **(3 marks)**

   (c) Find the gradient of C at each point where the curve meets the x-axis. **(2 marks)**

Tangents and normals

You can use **differentiation** to work out the equations of tangents and normals.

The curve drawn in black has equation

$y = x^2 - 2x + 4$

You can differentiate this to work out the gradient function:

$\dfrac{dy}{dx} = 2x - 2$

There is more about finding gradient functions on page 36.

At the point P (2, 4) the gradient of the curve is 2, so the gradient of the tangent is also 2.

The tangent passes through (2, 4) and it has equation $y = 2x$.

The normal is perpendicular to the tangent, so it has gradient $-\frac{1}{2}$.

The normal also passes through (2, 4) and it has equation $y = -\frac{1}{2}x + 5$.

There is more about parallel and perpendicular lines on page 18.

(graph) $y = x^2 - 2x + 4$

The **tangent** to the curve at P is a straight line that just touches the curve at P.

The **normal** to the curve at P is a straight line that is perpendicular to the tangent.

Golden rules

If a curve has gradient m at point P:

1 the **tangent** at P **also** has gradient m

2 the **normal** at P has gradient $-\dfrac{1}{m}$

Follow these steps:

1. Differentiate to find the gradient function.

2. Substitute $x = 2$ into $\dfrac{dy}{dx}$. The gradient at point P is -1.

3. Use $y - y_1 = m(x - x_1)$ to find the equation of a straight line with gradient -1 that passes through (2, 3).

There is more on finding equations of straight lines on page 17.

Worked example

The curve C has equation

$y = x^3 - 3x^2 - x + 9, \qquad x > 0$

The point P with coordinates (2, 3) lies on C.

Find the equation of the tangent to C at P, giving your answer in the form $y = mx + c$, where m and c are constants. **(5 marks)**

$\dfrac{dy}{dx} = 3x^2 - 6x - 1$

$= 3(2)^2 - 6(2) - 1$

$= 12 - 12 - 1$

$= -1$

$y - y_1 = m(x - x_1)$

$y - 3 = (-1)(x - 2)$

$y - 3 = -x + 2$

$y = -x + 5$

Now try this

The curve C has equation $y = f(x)$, $x > 0$, where

$\dfrac{dy}{dx} = \sqrt{x} + \dfrac{8}{x^2} - 5$

Given that the point P (4, 11) lies on C, find the equation of the normal to C at point P, giving your answer in the form $ax + by + c = 0$ **(4 marks)**

You have been given $\dfrac{dy}{dx}$, so start by substituting the x-coordinate of P to find the gradient of the **tangent** at P. The **normal** will be perpendicular to this.

Stationary points 1

You can use **calculus** to find the stationary points of a **graph** or **function** in your exam. You need to be confident with **differentiation** – have a look at pages 36 and 37 for a reminder.

Using differentiation

The stationary points of a graph or function are the points where the **derivative**, $\frac{dy}{dx}$ or f'(x), is equal to **zero**.

This graph has stationary points at P and Q. The slope of the curve is 0 at both points.

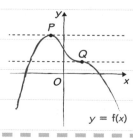

$y = f(x)$

Worked example

Find the coordinates of the stationary point on the curve with equation $y = 3x^2 + 12x + 5$

(4 marks)

$$\frac{dy}{dx} = 6x + 12$$

When $\frac{dy}{dx} = 0$, $6x + 12 = 0$

$6x = -12$

$x = -2$

So $y = 3 \times (-2)^2 + 12 \times (-2) + 5 = -7$

Stationary point is $(-2, -7)$.

To find the coordinates of the stationary point using calculus:

1. Differentiate to find $\frac{dy}{dx}$.
2. Set $\frac{dy}{dx} = 0$.
3. Solve the equation to find the value or values of x.
4. Find the corresponding value of y for each value of x.

Worked example

The diagram shows part of the curve with equation $y = 3x + \frac{12}{x^2} - 15$

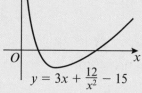

$y = 3x + \frac{12}{x^2} - 15$

Use calculus to show that y is increasing for $x > 2$

(4 marks)

$y = 3x + 12x^{-2} - 15$

$\frac{dy}{dx} = 3 - 24x^{-3} = 3 - \frac{24}{x^3}$

If $x > 2$ then $x^3 > 8$ and $\frac{24}{x^3} < 3$

So $3 - \frac{24}{x^3} > 0$

So if $x > 2$, $\frac{dy}{dx} \geq 0$ therefore y is increasing.

Increasing and decreasing functions

You can use the derivative to decide if a function is increasing or decreasing in a given interval:

- ✓ If f'(x) ⩾ 0 for $a < x < b$ then f(x) is INCREASING in the interval $a \leq x \leq b$.
- ✓ If f'(x) ⩽ 0 for $a < x < b$ then f(x) is DECREASING in the interval $a \leq x \leq b$.

The sign of f'(x) (+ or –) must be the same (or 0) in the whole interval, otherwise the function is neither decreasing nor increasing.

Now try this

1　Find the coordinates of the stationary point on the curve C with equation $y = x^2 - 8x + 3$
　(4 marks)

2　Use calculus to find the x-coordinates of the stationary points on the curve with equation $y = x^3 - 5x^2 + 8x + 1$　**(4 marks)**

You need to show that the **derivative** is **non-negative** for all values of x in the range given.

There are two stationary points. Differentiate, then solve a quadratic equation by factorising.

Stationary points 2

There are different types of stationary points on graphs. You need to be able to decide on the **nature** of a particular point. You can do this by finding the value of the **second derivative**, $\frac{d^2y}{dx^2}$ or $f''(x)$ at that point.

Maximum or minimum?

1 If $\frac{dy}{dx} = 0$ and $\frac{d^2y}{dx^2} < 0$ then the stationary point is a **maximum**.

At P, $\frac{dy}{dx} = 3 \times (-2)^2 - 12 = 0$ and $\frac{d^2y}{dx^2} = 6 \times (-2) = -12 < 0$ so P is a **maximum**.

2 If $\frac{dy}{dx} = 0$ and $\frac{d^2y}{dx^2} > 0$ then the stationary point is a **minimum**.

At Q, $\frac{dy}{dx} = 3 \times (2)^2 - 12 = 0$ and $\frac{d^2y}{dx^2} = 6 \times (2) = 12 > 0$ so Q is a **minimum**.

For a reminder about finding $\frac{d^2y}{dx^2}$ have a look at page 37.

$y = x^3 - 12x$

$\frac{dy}{dx} = 3x^2 - 12$ $\frac{d^2y}{dx^2} = 6x$

Worked example

The curve C has equation $y = 12\sqrt{x} - 2x$, $x > 0$

(a) Use calculus to find the coordinates of the turning point of C. **(4 marks)**

$y = 12x^{\frac{1}{2}} - 2x$

$\frac{dy}{dx} = 6x^{-\frac{1}{2}} - 2 = \frac{6}{\sqrt{x}} - 2$

When $\frac{dy}{dx} = 0$, $\frac{6}{\sqrt{x}} - 2 = 0$

$6 = 2\sqrt{x}$

$x = 9$

So $y = 12\sqrt{9} - 2 \times 9 = 18$

Turning point is $(9, 18)$.

(b) Find $\frac{d^2y}{dx^2}$ **(2 marks)**

$\frac{d^2y}{dx^2} = -3x^{-\frac{3}{2}}$

(c) State the nature of the turning point. **(1 mark)**

At turning point, $x = 9$, so $\frac{d^2y}{dx^2} = -3 \times 9^{-\frac{3}{2}}$

$= -\frac{1}{9}$

$\frac{d^2y}{dx^2} < 0$ so the turning point is a maximum.

Sketching gradient functions

You can use the features of $y = f(x)$ to sketch $y = f'(x)$.

Positive gradient, so $y = f'(x)$ is above the x-axis

Stationary point, so $y = f'(x)$ cuts the x-axis

Negative gradient so $y = f'(x)$ is below the x-axis

Vertical asymptotes in the same places on both graphs

$y = f(x)$

A turning point is a stationary point which is a **local maximum** or a **local minimum**.

For part (c), work out the value of $\frac{d^2y}{dx^2}$ when $x = 9$. Write down whether it is greater or less than 0 and state whether the turning point is a maximum or a minimum.

Now try this

The diagram shows a sketch of the curve with equation $y = 5x^2 - 3x - x^3$

The curve has stationary points at A and B.

(a) Use calculus to find the coordinates of A and B. **(6 marks)**

(b) Find the value of $\frac{d^2y}{dx^2}$ at B, and hence verify that B is a maximum. **(2 marks)**

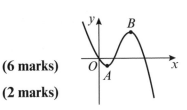

Modelling with calculus

You can use **calculus** to solve real-life problems involving maximums and minimums.

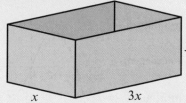

A cardboard box is made in the shape of an open-topped cuboid, with volume 18 000 cm³. The base of the cuboid has width x cm and length $3x$ cm. The height of the cuboid is y cm.

(a) Show that the area, A cm², of cardboard needed to make the box is given by

$$A = 3x^2 + \frac{48\,000}{x}$$ **(4 marks)**

Volume $= 3x^2y = 18\,000$

$y = \dfrac{6000}{x^2}$

$A = 3x^2 + 2 \times xy + 2 \times 3xy$

$\quad = 3x^2 + 8xy$

$\quad = 3x^2 + 8x\left(\dfrac{6000}{x^2}\right)$

$\quad = 3x^2 + \dfrac{48\,000}{x}$

(b) Use calculus to find the value of x for which A is stationary.   **(4 marks)**

$\dfrac{dA}{dx} = 6x - \dfrac{48\,000}{x^2}$

When $\dfrac{dA}{dx} = 0$, $6x - \dfrac{48\,000}{x^2} = 0$

$\qquad\qquad\qquad 6x^3 = 48\,000$

$\qquad\qquad\qquad x^3 = 8000$

$\qquad\qquad\qquad x = 20$

(c) Show that A is a minimum at this point and find its value.

    (4 marks)

$\dfrac{d^2A}{dx^2} = 6 + \dfrac{96\,000}{x^3}$

When $x = 20$, $\dfrac{d^2A}{dx^2} = 6 + \dfrac{96\,000}{20^3} = 18 > 0$

So A is a minimum.

$A = 3x^2 + \dfrac{48\,000}{x} = 3 \times (20)^2 + \dfrac{48\,000}{20}$

$\qquad\qquad = 3600$

This is the least area of carboard needed to make the box.

(a) Use the information given about the volume of the cuboid to write y in terms of x. Then write A in terms of x and y, and substitute your first expression to get A in terms of x only.

(b) Because A is a function of x only, you can differentiate with respect to x to find $\dfrac{dA}{dx}$. Find the stationary point of A by setting this equal to 0 and solving to find x.

(c) You need to find $\dfrac{d^2A}{dx^2}$ to determine the nature of the stationary point.

> You will need to use problem-solving skills throughout your exam – **be prepared!**

1 An oil well produces x barrels of oil each day. It models its profit, £P each day, using the formula $P = 80x - \dfrac{x^2}{50}$

(a) Find $\dfrac{dP}{dx}$ **(2 marks)**

(b) Hence show that P has a stationary point at $x = 2000$ and use calculus to determine the nature of that stationary point.   **(4 marks)**

2 The diagram shows a container in the shape of an open-topped cylinder, with height x m and radius r m. The cylinder has a capacity of 100 m³.

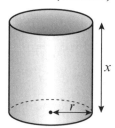

(a) Show that the area of sheet metal, A m², needed to make the tank is given by $A = \pi r^2 + \dfrac{200}{r}$ **(4 marks)**

(b) Use calculus to find the value of r for which A is stationary.   **(4 marks)**

(c) Prove that this value of r gives a minimum value of A.   **(2 marks)**

(d) Hence calculate the minimum area of sheet metal needed to make the tank.   **(2 marks)**

When you differentiate, remember that π is constant

Integration

Integration is the **opposite** of differentiation. You can use this rule to integrate terms which are written in the form ax^n.

You **increase** the power by 1 ...

This is the symbol for integration.

... then **divide** by the **new** power.

This rule **doesn't** work if the original power is −1

$$\int x^n dx = \frac{x^{n+1}}{n+1} + c, \, n \neq -1$$

You are integrating with **respect** to x.

You have to add the **constant of integration**.

To **integrate** a function, write each term in the form ax^n, then integrate one term at a time.

The constant of integration

When you **differentiate**, any **constant terms** disappear. So lots of functions have the same derivative.

$y = x^2 + 5$

$y = x^2$

Differentiate $\frac{dy}{dx} = 2x$ Integrate $y = x^2 + c$

$y = x^2 - 19$

When you integrate you don't know the constant. You write '+ c' at the end to show this. This is called **indefinite integration**.

Worked example

Find $\int(12x^3 + 6x - 15x^{\frac{2}{3}}) \, dx$, giving each term in its simplest form. **(5 marks)**

$$\int\left(12x^3 + 6x - 15x^{\frac{2}{3}}\right) dx$$

$$= \frac{12x^4}{4} + \frac{6x^2}{2} - \frac{15x^{\frac{5}{3}}}{\left(\frac{5}{3}\right)} + c$$

$$= 3x^4 + 3x^2 - 9x^{\frac{5}{3}} + c$$

Integrate term-by-term and don't forget to add the constant of integration. For each term:
- increase the power by 1
- divide by the **new** power.

$\frac{2}{3} + 1 = \frac{5}{3}$. Dividing by $\frac{5}{3}$ is the same as dividing by 5 then multiplying by 3.

Golden rules

1 Write every term in a polynomial in the form ax^n **before** integrating.

2 Remember to include the **constant of integration**.

3 Simplify any **coefficients** if possible.

Worked example

Given that $y = \frac{1}{x^3} - 3x^5, \, x \neq 0$, find $\int y \, dx$ **(3 marks)**

$$\int(x^{-3} - 3x^5) \, dx = \frac{x^{-2}}{-2} - \frac{3x^6}{6} + c$$

$$= -\frac{1}{2}x^{-2} - \frac{1}{2}x^6 + c$$

Be careful with negative powers. For the first term, you have to **increase** the power of −3 by 1 to get −2, then divide by the **new power**, −2.

Now try this

1 Find $\int(1 - 3x^3) \, dx$ **(3 marks)**

2 Find $\int(3x + 1)^2 \, dx$ **(4 marks)**

3 Given that $y = 6x^2 + 5x\sqrt{x}, \, x > 0$, find $\int y \, dx$ **(3 marks)**

Expand the brackets first.

42

Finding the constant

If you know **one point** on the original curve, or **one value** of f(x), then you can calculate the constant of integration.

Using substitution

All three of these curves have the same gradient function:

$$\frac{dy}{dx} = 2x$$

But only one passes through the point $(2, -1)$. You can find its equation by integrating, then substituting $x = 2$ and $y = -1$ to find the value of c.

$$y = x^2 + c$$
$$-1 = 2^2 + c$$
$$c = -5$$
$$y = x^2 - 5$$

Graph showing $y = x^2 + 4$, $y = x^2$, $y = x^2 - 5$ with point $(2, -1)$

Worked example

$$\frac{dy}{dx} = 7 + \frac{1}{\sqrt{x}}, \, x > 0$$

Given that $y = 26$ at $x = 4$, find y in terms of x. **(6 marks)**

$$\frac{dy}{dx} = 7 + x^{-\frac{1}{2}}$$

$$y = 7x + \frac{x^{\frac{1}{2}}}{\left(\frac{1}{2}\right)} + c$$

$$= 7x + 2x^{\frac{1}{2}} + c$$

$$26 = 7(4) + 2(4)^{\frac{1}{2}} + c$$

$$= 28 + 4 + c$$

$$c = -6$$

$$y = 7x + 2x^{\frac{1}{2}} - 6$$

Worked example

Given that f$(-1) = 9$ and f$'(x) = 6x^2 - 10x - 3$, find f(x). **(5 marks)**

$$f(x) = \frac{6x^3}{3} - \frac{10x^2}{2} - 3x + c$$

$$= 2x^3 - 5x^2 - 3x + c$$

$$f(-1) = 2(-1)^3 - 5(-1)^2 - 3(-1) + c$$

$$= -2 - 5 + 3 + c$$

$$= -4 + c$$

$$9 = -4 + c$$

$$c = 13$$

$$f(x) = 2x^3 - 5x^2 - 3x + 13$$

f$(-1) = 9$ is the same as saying that the curve with equation $y =$ f(x) passes through the point $(-1, 9)$.

Three key steps

1 Integrate, and remember to include the constant of integration.

2 Substitute the values of x and y you know and solve an equation to find c.

3 Write out the function including the constant of integration you've found.

---

1 The curve C has equation $y =$ f(x), $x > 0$, and the point $(4, 17)$ lies on C.

Given that f$'(x) = 3 - \dfrac{2 + 3\sqrt{x}}{x^2}$, find f$(x)$. **(5 marks)**

2 $\dfrac{dy}{dx} = \dfrac{(x^2 + 5)^2}{x^2}, \, x \neq 0$

(a) Show that $\dfrac{dy}{dx} = x^2 + 10 + 25x^{-2}$ **(2 marks)**

(b) Given that $y = -13$ at $x = 1$, find y in terms of x. **(6 marks)**

Definite integration

In your exam, you might have to find an integral with **limits**. This is called definite integration. You should make sure you are confident with **indefinite integration** before revising this – check page 42 for a reminder.

Evaluating a definite integral

A definite integral has a **numerical answer**.

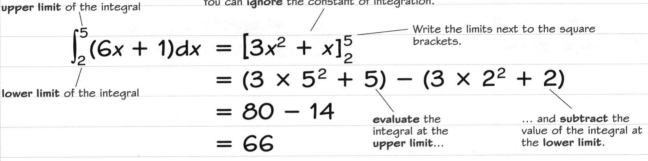

upper limit of the integral

Integrate (6x + 1) in the normal way and write the integral in **square brackets**. You can **ignore** the constant of integration.

Write the limits next to the square brackets.

$$\int_2^5 (6x + 1)\,dx = [3x^2 + x]_2^5$$

lower limit of the integral

$$= (3 \times 5^2 + 5) - (3 \times 2^2 + 2)$$

$$= 80 - 14$$

$$= 66$$

evaluate the integral at the upper limit...

... and **subtract** the value of the integral at the lower limit.

Use calculus to find the value of

$$\int_1^9 (2x + 6\sqrt{x})\,dx \qquad \textbf{(5 marks)}$$

$$\int_1^9 \left(2x + 6x^{\frac{1}{2}}\right)dx = \left[x^2 + 4x^{\frac{3}{2}}\right]_1^9$$

$$= \left(9^2 + 4 \times 9^{\frac{3}{2}}\right) - \left(1^2 + 4 \times 1^{\frac{3}{2}}\right)$$

$$= 189 - 5$$

$$= 184$$

Some calculators have a key like the one shown here which can work out numerical integration.

You **cannot** just use this key and write down your answer. You need to **use calculus** and **show your working**, or you will get a maximum of 1 mark. You could use this key to check your answer, but it's safer to stay away from it entirely!.

If you are using calculus to do definite integration you will usually need to give an exact answer. If your answer isn't a whole number or a fraction, write it in simplified **surd form**.

$$2\sqrt{12} = 2\sqrt{4 \times 3} = 2\sqrt{4}\sqrt{3} = 4\sqrt{3}$$

Have a look at page 3 for a reminder about surds.

Find the exact value of $\int_1^{12} \left(\dfrac{1}{\sqrt{x}}\right)dx$ **(4 marks)**

$$\int_1^{12} \left(x^{-\frac{1}{2}}\right)dx = \left[2x^{\frac{1}{2}}\right]_1^{12}$$

$$= (2\sqrt{12}) - (2\sqrt{1})$$

$$= 4\sqrt{3} - 2$$

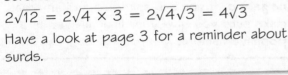

Start by writing $\dfrac{6}{x^2}$ as $6x^{-2}$.

1 Use calculus to find the exact value of

$$\int_1^3 \left(3x^2 - 7 + \dfrac{6}{x^2}\right)dx \qquad \textbf{(5 marks)}$$

2 Given that $f(x) = \dfrac{6}{x^3} - \dfrac{2}{\sqrt{x}}$, find $\int_1^2 f(x)\,dx$, giving your answer in the form $a - b\sqrt{2}$, where a and b are constants. **(5 marks)**

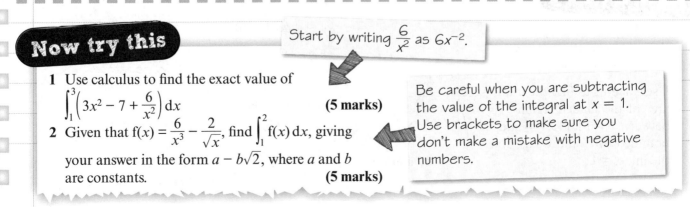

Be careful when you are subtracting the value of the integral at $x = 1$. Use brackets to make sure you don't make a mistake with negative numbers.

Area under a curve

You can use **definite integration** to find the area between a curve and the x-axis. The area between the curve $y = f(x)$, the x-axis and the lines $x = a$ and $x = b$ is given by

$$A = \int_a^b f(x)\, dx$$

Look at the previous page for a reminder about definite integration.

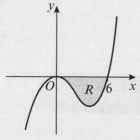

Worked example

The diagram shows part of the curve with equation $y = (x + 3)(1 - x)$

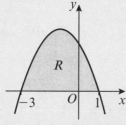

Use calculus to find the exact area of the shaded region, R. **(5 marks)**

$y = 3 - 2x - x^2$

$$\int_{-3}^{1}(3 - 2x - x^2)\, dx = \left[3x - x^2 - \frac{x^3}{3}\right]_{-3}^{1}$$

$$= (3 - 1 - \tfrac{1}{3}) - (-9 - 9 + 9)$$

$$= \frac{5}{3} - (-9) = 10\frac{2}{3}$$

Area of $R = 10\frac{2}{3}$

The curve crosses the x-axis at -3 and 1. So the **limits** for your **definite integral** will be -3 and 1. Always put the **right-hand** boundary as the **upper limit** and the left-hand boundary as the **lower limit**. You can give your exact answer as a fraction, mixed number or decimal.

Worked example

The diagram shows part of the curve with equation $y = x^2(x - 6)$

Find the area of the shaded region, R. **(6 marks)**

$y = x^3 - 6x^2$

$$\int_0^6 (x^3 - 6x^2)\, dx = \left[\frac{x^4}{4} - 2x^3\right]_0^6$$

$$= (324 - 432) - (0 - 0)$$

$$= -108$$

Area of $R = 108$

Negative areas

When you use a **definite integral** to find an area **below** the x-axis, the answer will be **negative**. If you are asked to find an area, make sure you give your final answer as a **positive** number.

Now try this

The diagram shows part of the curve C with equation $y = x(x - 2)(x - 4)$

Use calculus to find the total area of the shaded region, between $x = 1$ and $x = 4$ and bounded by C, the x-axis, and the line $x = 1$ **(9 marks)**

Be careful with this question – you can't just find $\int_1^4 y\, dx$ because an area **below** the x-axis will produce a **negative** integral. You need to work out **two separate** definite integrals to find these two areas, then add the areas together:

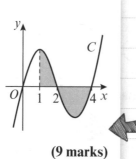

$$A_1 = \int_1^2 y\, dx$$

$$A_2 = -\int_2^4 y\, dx$$

Total area $= A_1 + A_2$

More areas

You can use areas of **triangles** and **trapeziums**, together with **definite integration**, to find areas enclosed by curves and straight lines. Here are three examples.

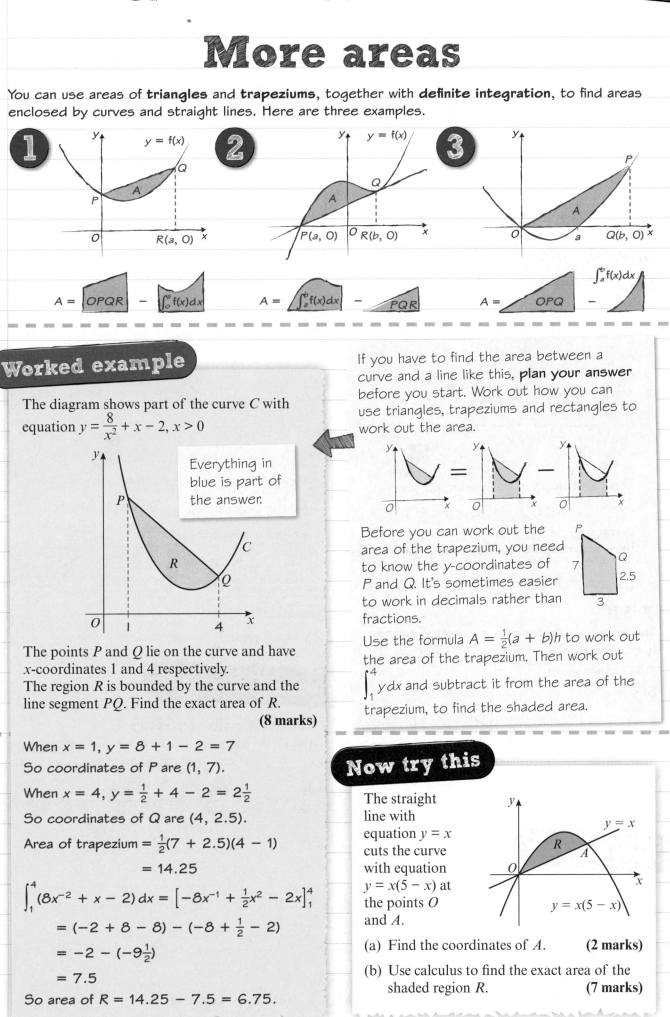

1 $y = f(x)$, Q, A, P, O, $R(a, 0)$ x

2 $y = f(x)$, Q, A, $P(a, 0)$, O, $R(b, 0)$ x

3 P, A, O, a, $Q(b, 0)$ x

$A = \boxed{OPQR} - \int_0^a f(x)\,dx$

$A = \int_a^b f(x)\,dx - \boxed{PQR}$

$A = \boxed{OPQ} - \int_a^b f(x)\,dx$

Worked example

The diagram shows part of the curve C with equation $y = \dfrac{8}{x^2} + x - 2$, $x > 0$

Everything in blue is part of the answer.

The points P and Q lie on the curve and have x-coordinates 1 and 4 respectively.
The region R is bounded by the curve and the line segment PQ. Find the exact area of R.

(8 marks)

When $x = 1$, $y = 8 + 1 - 2 = 7$

So coordinates of P are $(1, 7)$.

When $x = 4$, $y = \frac{1}{2} + 4 - 2 = 2\frac{1}{2}$

So coordinates of Q are $(4, 2.5)$.

Area of trapezium $= \frac{1}{2}(7 + 2.5)(4 - 1)$

$\qquad = 14.25$

$\displaystyle\int_1^4 (8x^{-2} + x - 2)\,dx = \left[-8x^{-1} + \frac{1}{2}x^2 - 2x\right]_1^4$

$\qquad = (-2 + 8 - 8) - (-8 + \frac{1}{2} - 2)$

$\qquad = -2 - (-9\frac{1}{2})$

$\qquad = 7.5$

So area of $R = 14.25 - 7.5 = 6.75$.

If you have to find the area between a curve and a line like this, **plan your answer** before you start. Work out how you can use triangles, trapeziums and rectangles to work out the area.

Before you can work out the area of the trapezium, you need to know the y-coordinates of P and Q. It's sometimes easier to work in decimals rather than fractions.

Use the formula $A = \frac{1}{2}(a + b)h$ to work out the area of the trapezium. Then work out $\displaystyle\int_1^4 y\,dx$ and subtract it from the area of the trapezium, to find the shaded area.

Now try this

The straight line with equation $y = x$ cuts the curve with equation $y = x(5 - x)$ at the points O and A.

(a) Find the coordinates of A. **(2 marks)**

(b) Use calculus to find the exact area of the shaded region R. **(7 marks)**

Exponential functions

You need to be able to sketch the graph of $y = a^x$. You can only sketch this graph when a is a **positive number**.

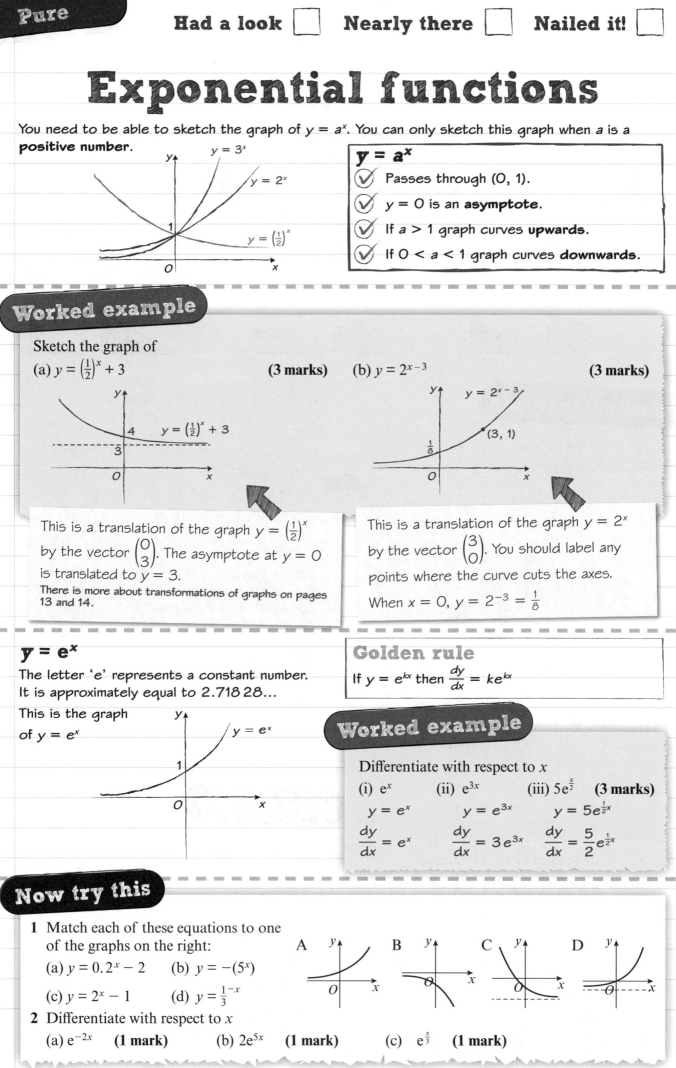

$y = 3^x$

$y = 2^x$

$y = \left(\frac{1}{2}\right)^x$

$y = a^x$
- ✓ Passes through (0, 1).
- ✓ $y = 0$ is an **asymptote**.
- ✓ If $a > 1$ graph curves **upwards**.
- ✓ If $0 < a < 1$ graph curves **downwards**.

Worked example

Sketch the graph of

(a) $y = \left(\frac{1}{2}\right)^x + 3$ **(3 marks)**

(b) $y = 2^{x-3}$ **(3 marks)**

$y = \left(\frac{1}{2}\right)^x + 3$

$y = 2^{x-3}$

(3, 1)

$\frac{1}{8}$

This is a translation of the graph $y = \left(\frac{1}{2}\right)^x$ by the vector $\begin{pmatrix} 0 \\ 3 \end{pmatrix}$. The asymptote at $y = 0$ is translated to $y = 3$.

There is more about transformations of graphs on pages 13 and 14.

This is a translation of the graph $y = 2^x$ by the vector $\begin{pmatrix} 3 \\ 0 \end{pmatrix}$. You should label any points where the curve cuts the axes.

When $x = 0$, $y = 2^{-3} = \frac{1}{8}$

$y = e^x$

The letter 'e' represents a constant number. It is approximately equal to $2.718\,28\ldots$

This is the graph of $y = e^x$

$y = e^x$

Golden rule

If $y = e^{kx}$ then $\frac{dy}{dx} = ke^{kx}$

Worked example

Differentiate with respect to x

(i) e^x (ii) e^{3x} (iii) $5e^{\frac{x}{2}}$ **(3 marks)**

$y = e^x$ $y = e^{3x}$ $y = 5e^{\frac{1}{2}x}$

$\frac{dy}{dx} = e^x$ $\frac{dy}{dx} = 3e^{3x}$ $\frac{dy}{dx} = \frac{5}{2}e^{\frac{1}{2}x}$

Now try this

1 Match each of these equations to one of the graphs on the right:

  (a) $y = 0.2^x - 2$ (b) $y = -(5^x)$

  (c) $y = 2^x - 1$ (d) $y = \frac{1}{3}^{-x}$

A   B   C   D

2 Differentiate with respect to x

  (a) e^{-2x} **(1 mark)**   (b) $2e^{5x}$ **(1 mark)**   (c) $e^{\frac{x}{3}}$ **(1 mark)**

Logarithms

Logarithms (or **logs**) are a way of writing facts about **powers**. These two statements mean the same thing:

You say 'log to the base a of b equals x'.

$$\log_a b = x \longleftrightarrow a^x = b$$

a is the **base** of the logarithm.

For example: $\log_3 9 = 2 \longleftrightarrow 3^2 = 9$

Remembering the order

The key to being confident in log questions is remembering the **basic definition**.
Start at the **base**, and work in a **circle**.

... to get b

$$\log_a b = x$$

Raise a to the power x...

Laws of logarithms

Learn these four key laws for manipulating expressions involving logs. These laws all work for logarithms with **the same base**.

1 $\log_a x + \log_a y = \log_a (xy)$

$\log_4 8 + \log_4 2 = \log_4 16 = 2$ (since $4^2 = 16$)

2 $\log_a x - \log_a y = \log_a \left(\dfrac{x}{y}\right)$

$\log_9 18 - \log_9 6 = \log_9 3 = \frac{1}{2}$ (since $9^{\frac{1}{2}} = 3$)

3 $\log_a \left(\dfrac{1}{x}\right) = -\log_a x$

$\log_8 \left(\frac{1}{2}\right) = -\log_8 2 = -\frac{1}{3}$

4 $\log_a (x^n) = n\log_a x$

$\log_5 (25^3) = 3\log_5 25 = 3 \times 2 = 6$

Worked example

Find
(a) the positive value of x such that
$\log_x 49 = 2$ **(2 marks)**

$x^2 = 49$

$x = 7$

(b) the value of y such that $\log_5 y = -2$ **(2 marks)**

$5^{-2} = y$

$y = \frac{1}{25}$

Write down the corresponding power fact. Remember:
$\log_a b = x \longleftrightarrow a^x = b$

Worked example

Express $3\log_a 2 + \log_a 10$ as a single logarithm to base a.   **(3 marks)**

$3\log_a 2 + \log_a 10 = \log_a(2^3) + \log_a 10$
$$= \log_a(2^3 \times 10)$$
$$= \log_a 80$$

Use law 4 to write $3\log_a 2$ as $\log_a(2^3)$, then use law 1 to combine the two logarithms.

Special cases

Learn these two special cases to save time:

1 $\log_a a = 1$ **2** $\log_a 1 = 0$

Now try this

1 Find
   (a) the value of y such that $\log_3 y = -1$ **(2 marks)**
   (b) the value of p such that $\log_p 8 = 3$ **(2 marks)**
   (c) the value of $\log_4 8$ **(2 marks)**

2 Express as a single logarithm to base a
   (a) $2\log_a 5$ **(2 marks)**
   (b) $\log_a 2 + \log_a 9$ **(2 marks)**
   (c) $3\log_a 4 - \log_a 8$ **(3 marks)**

3 Show that $2\log_8 6 - \log_8 9 = \frac{2}{3}$

Equations with logs

If you see an equation involving **logarithms** in your exam, you will probably need to rearrange it using the **laws of logarithms**, which are covered on page 48.

Two steps to solving log equations

Follow these two steps to solve most log equations in your exam:

 Group the log terms on one side, then use the laws of logs on page 48 to write them as a **single algorithm**.

 Rewrite $\log_a f(x) = k$ as $f(x) = a^k$ and solve the equation to find x.

Undefined logs

The value $\log_a b$ is **only defined** for $b > 0$. You can't calculate $\log_a 0$ or the log of any negative number. If an equation contains $\log_a x$ or $\log_a kx$ then **ignore** any solutions where $x \leqslant 0$.

If there are solutions to ignore in an exam question, you will usually be given a **range** of possible values for x.

Worked example

Solve the equation $2\log_5 x - \log_5 3x = 2$ **(4 marks)**

$\log_5 x^2 - \log_5 3x = 2$

$\log_5 \left(\dfrac{x^2}{3x} \right) = 2$

$\dfrac{x^2}{3x} = 5^2 = 25$

$x^2 = 75x$

$x^2 - 75x = 0$

$x(x - 75) = 0$ so $\cancel{x = 0}$ or $\underline{x = 75}$

Follow the two steps given above:

1. Rearrange the left-hand side into a single logarithm.

2. Write the corresponding power fact:
$$\log_5 f(x) = 2 \rightarrow f(x) = 5^2$$

You need to solve two **simultaneous equations**. You can ignore the negative square root because p and q are positive.

Worked example

Given that $0 < x < 2$ and $\log_3 (2 - x) - 2\log_3 x = 1$, find the value of x.   **(6 marks)**

$\log_3 (2 - x) - \log_3 (x^2) = 1$

$\log_3 \left(\dfrac{2 - x}{x^2} \right) = 1$

$\dfrac{2 - x}{x^2} = 3$

$2 - x = 3x^2$

$3x^2 + x - 2 = 0$

$(3x - 2)(x + 1) = 0$

$x = \dfrac{2}{3}$ $x = \cancel{-1}$

Ignore this solution because $0 < x < 2$.

Worked example

p and q are positive constants, with
$$p = 5q \quad \text{①}$$
$$\log_5 p + \log_5 q = 2 \quad \text{②}$$

Find the exact values of p and q.   **(6 marks)**

From ②: $\log_5 (pq) = 2$

Substituting ①: $\log_5 (5q^2) = 2$

$5q^2 = 5^2 = 25$

$q^2 = 5$

$q = \sqrt{5}$

Substituting into ①: $p = 5\sqrt{5}$

Now try this

1 Solve $\log_2 (x + 1) - \log_2 x = \log_2 5$ **(3 marks)**

2 Solve the equation $\log_6 (x - 1) + \log_6 x = 1$ **(4 marks)**

3 Solve $\log_3 (x - 1) = -1$ **(2 marks)**

4 Find the values of x such that $2\log_4 x - \log_4 (x - 3) = 2$ **(5 marks)**

Exponential equations

You can find unknown powers in equations using the log functions on your **calculator**. Make sure you **write down** any logarithms you are working out.

 You can use this key to work out logs to any base.

 This key means $\log_{10}$. You can **take logs** of both sides of an equation and solve it using this key.

Worked example

(a) Solve the equation $4^x = 13$, giving your answer to 3 significant figures. **(3 marks)**

$x = \log_4 13 = 1.85$ (3 s.f.)

(b) Find, to 3 significant figures, the value of y for which $5^y = 4$ **(3 marks)**

$\log(5^y) = \log 4$

$y \log 5 = \log 4$

$y = \dfrac{\log 4}{\log 5} = 0.861$ (3 s.f.)

Make sure you **write down** the logarithm you need to find even if you are working it out on your calculator in one go. Part (a) shows a method using the log▪☐ key. Part (b) shows a method by **taking logs** of both sides and using the laws of logs. This works for any base, so you can use the log key on your calculator.

Use the fact that $5^{2x} = (5^x)^2$ to write a **quadratic equation** in 5^x. For a reminder on the **laws of indices** have a look at page 1. It might help to write the equation as $Y^2 - 3Y + 2 = 0$, with $Y = 5^x$.

Factorising gives you two values for 5^x. Each of these gives you a value for x. Remember that $\log_a 1 = 0$ for any base, so $\log_5 1 = 0$.

Worked example

Solve the equation

$5^{2x} - 3(5^x) + 2 = 0$

giving your answers to 2 decimal places where appropriate. **(6 marks)**

$(5^x)^2 - 3(5^x) + 2 = 0$

$(5^x - 2)(5^x - 1) = 0$

$5^x = 2$ $5^x = 1$

$x = \log_5 2 = 0.43$ (2 d.p.) $x = \log_5 1 = 0$

Worked example

Solve the equation $2^x = 5^{x-3}$, giving your answer to 3 significant figures. **(4 marks)**

$\log 2^x = \log 5^{x-3}$

$x \log 2 = (x - 3) \log 5$

$3 \log 5 = x \log 5 - x \log 2$

$3 \log 5 = x(\log 5 - \log 2)$

$x = \dfrac{3 \log 5}{\log 5 - \log 2} = 5.27$ (3 s.f.)

Problem solved!

This exponential equation has different bases on each side. You can still take logs of both sides but you must take logs **to the same base** on each side.

You will need to use problem-solving skills throughout your exam – **be prepared!** 💡

Now try this

1 Find, to 3 significant figures
  (a) the value of b for which $2^b = 15$ **(3 marks)**
  (b) the value of x for which $6^x = 0.4$ **(3 marks)**

2 (a) Solve the equation $3^{2x} + 3^x = 6$, giving your answer to 2 decimal places. **(6 marks)**
  (b) Explain why there is only one solution to the equation $3^{2x} + 3^x = 6$ **(1 mark)**

3 Solve $3^{x-1} = 2^{2x+1}$, giving your answer correct to 3 significant figures. **(4 marks)**

4 $f(x) = 6^{x^2}, x \in \mathbb{R}$ $g(x) = 2^{x-1}, x \in \mathbb{R}$

Show that the curves $y = f(x)$ and $y = g(x)$ do not intersect. **(5 marks)**

Set $6^{x^2} = 2^{x-1}$ and use the discriminant of the resulting quadratic equation in x.

Natural logarithms

A logarithm to the base e is sometimes called a natural logarithm. The graph of $y = \ln x$ is the graph of $y = e^x$ reflected in the line $y = x$. It has an asymptote at $x = 0$ and crosses the x-axis at (1, 0).

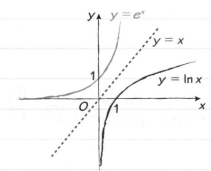

Golden rule

$\ln x$ works just like any other logarithm: $e^{\ln x} = \ln(e^x) = x$

Look for the $\boxed{e^{\blacksquare}}$ and $\boxed{\ln}$ functions on your calculator to find values of e^x and $\ln x$.

Worked example

Find the exact solutions to the equation

$e^x + 3e^{-x} = 4$ **(4 marks)**

$(e^x)^2 + 3 = 4e^x$

$(e^x)^2 - 4e^x + 3 = 0$

$(e^x - 3)(e^x - 1) = 0$

$e^x = 3$ $e^x = 1$

$x = \ln 3$ $x = \ln 1 = 0$

You can write this as a quadratic equation using the substitution $u = e^x$:

$$u + \frac{3}{u} = 4$$

$u^2 - 4u + 3 = 0$

If a question asks for exact solutions then you should leave your answers as logs, or powers of e, and simplify them as much as possible.

Problem solved!

The laws of logarithms work exactly the same with ln as they do with $\log_{10}$ and $\log_a$

You can't combine the $2x$ with the e^{3x+1} easily, so just take logs of both sides. You can then use the laws of logs to simplify the left-hand side. Group the x terms together, then factorise to get x on its own.

You will need to use problem-solving skills throughout your exam – **be prepared!**

Worked example

Solve $2^x e^{3x+1} = 10$, giving your answer in the form $\dfrac{a + \ln b}{c + \ln d}$ where a, b, c and d are integers to be found. **(5 marks)**

$\ln(2^x e^{3x+1}) = \ln 10$

$\ln 2^x + \ln(e^{3x+1}) = \ln 10$

$x\ln 2 + 3x + 1 = \ln 10$

$x(\ln 2 + 3) = \ln 10 - 1$

$$x = \frac{-1 + \ln 10}{3 + \ln 2}$$

Now try this

1 Given that $f(x) = \ln x$, $x > 0$, sketch, on separate axes, the graphs of
  (a) $y = f(x - 2)$ **(2 marks)**
  (b) $y = -f(x)$ **(2 marks)**
  (c) $y = f(3x)$ **(2 marks)**

2 The point P with y-coordinate 6 lies on the curve with equation $y = 3e^{2x-1}$. Find, in terms of $\ln 2$, the x-coordinate of P. **(2 marks)**

3 Solve
  (a) $\ln(x + 1) - \ln x = \ln 5$ **(2 marks)**
  (b) $e^{4x} + 3e^{2x} = 10$ **(5 marks)**
  (c) $\ln(6x + 7) = 2\ln x$, $x > 0$ **(4 marks)**

4 Find the exact solution to the equation
  $3^x e^{2x-5} = 7$ **(5 marks)**

5 The function f is defined by
$$f : x \mapsto \frac{3x^2 - 7x + 2}{x^2 - 4}, \quad x \neq \pm 2$$
  (a) Show that $f(x) = \dfrac{3x - 1}{x + 2}$ **(3 marks)**
  (b) Hence, or otherwise, solve the equation
    $\ln(3x^2 - 7x + 2) = 1 + \ln(x^2 - 4)$, $x > 2$
    giving your answer in terms of e.
                       (4 marks)

Start with $6 = 3e^{2x-1}$. Divide both sides by 3 then take the natural logarithm of both sides.

Exponential modelling

You can use exponential functions to model lots of real-life situations.

1 Growth models

A typical growth model can be described as $N = C + N_0 e^{kt}$

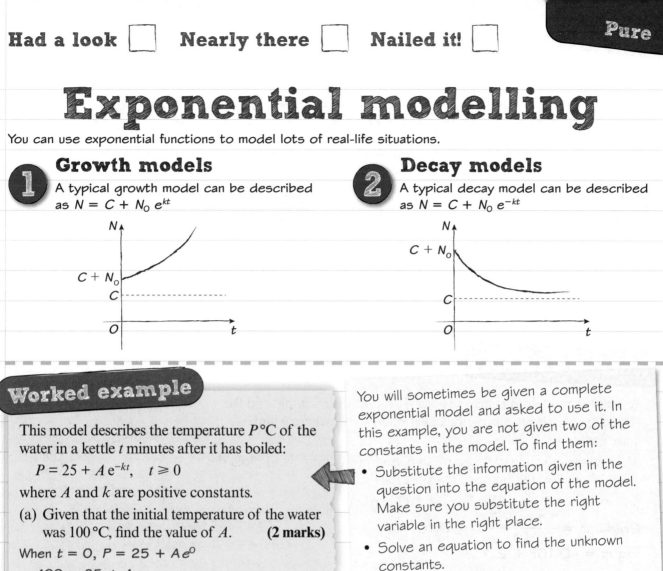

2 Decay models

A typical decay model can be described as $N = C + N_0 e^{-kt}$

Worked example

This model describes the temperature $P\,°C$ of the water in a kettle t minutes after it has boiled:

$$P = 25 + A e^{-kt}, \quad t \geq 0$$

where A and k are positive constants.

(a) Given that the initial temperature of the water was $100\,°C$, find the value of A. **(2 marks)**

When $t = 0$, $P = 25 + A e^0$

$\quad 100 = 25 + A$

$\qquad A = 75$

After 10 minutes, the water in the kettle has cooled down to $40\,°C$.

(b) Show that $k = \frac{1}{10} \ln 5$. **(3 marks)**

$\quad 40 = 25 + 75 e^{-10k}$

$\quad 15 = 75 e^{-10k}$

$\quad \frac{1}{5} = e^{-10k}$

$-10k = \ln\left(\frac{1}{5}\right) = \ln(5^{-1}) = -\ln 5$

So $k = \frac{1}{10} \ln 5$

(c) Find the temperature of the water after 18 minutes, in $°C$ to 1 decimal place. **(2 marks)**

$P = 25 + 75 e^{-\left(\frac{1}{10}\ln 5\right) \times 18} = 29.1\,°C$ (1 d.p.)

You will sometimes be given a complete exponential model and asked to use it. In this example, you are not given two of the constants in the model. To find them:

- Substitute the information given in the question into the equation of the model. Make sure you substitute the right variable in the right place.

- Solve an equation to find the unknown constants.

Rates of change

You might have to find the **rate of change** of a quantity in an exponential model. You can do this by **differentiating** with respect to time. In the model in the Worked example on the left, the rate of change of temperature with time would be given by

$$\frac{dP}{dt} = -kA e^{-kt}$$

The units would be $°C$/min. The fact that the rate of change is **negative** tells you that the temperature is **decreasing**.

Now try this

1 The number of cells, N, in a bacterial culture at a time t hours after midday is modelled as

$$N = 100 e^{0.8t}, \quad t \geq 0$$

(a) Write down the number of cells in the culture at midday. **(1 mark)**

(b) Show that the rate of change of N at time t is directly proportional to the size of the population. **(3 marks)**

2 The mass, M grams, of a sample of radon after t hours is modelled using the equation

$$M = 250 e^{-kt}, \quad t \geq 0$$

where k is a positive constant.

(a) What was the initial mass of the sample? **(1 mark)**

After 90 hours the sample has lost half its mass.

(b) Find the value of k to 3 significant figures. **(4 marks)**

Modelling with logs

You can use logarithms to determine the constants in some **exponential** and **polynomial** models. There are two different cases that you need to know about:

① The polynomial model
$$y = ax^n$$
In this model a and n are constants.
If x and y satisfy this model then the graph of **log y against log x** will be a straight line: $\log y = n \log x + \log a$

gradient = n y-intercept = $\log a$

② The exponential model
$$y = kb^x$$
In this model k and b are constants.
If x and y satisfy this model then the graph of **log y against x** will be a straight line: $\log y = (\log b) x + \log a$

gradient = $\log b$ y-intercept = $\log a$

Worked example

A scientist models the mass, m grams, of a fluorine sample and the time elapsed, t hours, using the equation $m = pq^t$, where p and q are constants.

She observes the actual mass over a period of 10 hours, and plots the graph shown on the right, of t against $\log m$.

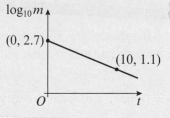

(a) Find an equation for the line.   **(2 marks)**

$$\text{Gradient} = \frac{1.1 - 2.7}{10} = -0.16$$

$$\log m = -0.16t + 2.7$$

> For part (b), the safest approach is to rearrange the equation of the line into the form $m = pq^t$, then compare values.
>
> **Check it!**
>
> $501 \times 0.692^{10} = 12.6$
>
> $\log 12.6 = 1.1$ ✓

(b) Determine the values of p and q in the model to 3 significant figures.   **(4 marks)**

$$m = 10^{-0.16t + 2.7}$$
$$= 10^{2.7} \times (10^{-0.16})^t$$
$$= 501 \times 0.692^t$$

So $p = 501$ and $q = 0.692$ (3 s.f.)

(c) Interpret these values in the context of the model.   **(2 marks)**

$p = 501$ is the initial mass of the sample in grams.

$q = 0.692$ is the proportional change in the sample each hour.

(d) Use the model to predict the mass of the sample after 3 days.   **(1 mark)**

3 days = 72 hours, so $m = 501 \times 0.692^{72} = 1.51 \times 10^9$ g

(e) Give one reason why this prediction may not be accurate.   **(1 mark)**

The model is based on 10 hours of data, so it may not be accurate over a longer period.

Now try this

A computer algorithm is used to allocate medical students to hospitals. When there are N students, the runtime of the algorithm, x milliseconds, is expected to follow the rule $x = aN^b$, where a and b are constants.

(a) Show that this relationship can be written in the form $\log x = k \log N + c$, giving k and c in terms of a and b.   **(2 marks)**

The algorithm is run a number of times and the following values of x and N are found:

N	1000	1500	2000	2500	3000	3500	4000
x	460	980	1660	2510	3520	4680	5990

(b) Plot a graph of $\log x$ against $\log N$, and comment on the accuracy of the expected model.   **(3 marks)**

(c) Find the values of a and b, giving your answers to 2 significant figures.   **(4 marks)**

You are the examiner!

In your exam you might be asked to identify errors in working. You should also be confident **checking your own work**. Each of these students has made a key mistake in their working. Can you spot them all?

1 Simplify $\dfrac{5 - 2\sqrt{3}}{\sqrt{3} - 1}$ giving your answer in the form $p + q\sqrt{3}$, where p and q are rational numbers. **(4 marks)**

$$\dfrac{5 - 2\sqrt{3}}{\sqrt{3} - 1} \times \dfrac{\sqrt{3} + 1}{\sqrt{3} + 1}$$

$$= \dfrac{5 - 2\sqrt{3}(\sqrt{3} + 1)}{(\sqrt{3})^2 - \sqrt{3} + \sqrt{3} - 1}$$

$$= \dfrac{5 - 6 - 2\sqrt{3}}{2}$$

$$= -\dfrac{1}{2} - \sqrt{3}$$

2 The equation $x^2 + 3px + p = 0$, where p is a non-zero constant, has equal roots. Find the value of p. **(4 marks)**

$$b^2 - 4ac = 0$$
$$3p^2 - 4p = 0$$
$$p(3p - 4) = 0$$
$$\cancel{p = 0} \text{ or } p = \dfrac{4}{3}$$

3 Find $\int(12x^5 - 8x^3 + 3)\,\mathrm{d}x$, giving each term in its simplest form. **(4 marks)**

$$\int(12x^5 - 8x^3 + 3)\,\mathrm{d}x = \dfrac{12x^6}{6} - \dfrac{8x^4}{4} + \dfrac{3x}{1}$$

$$= 2x^6 - 2x^4 + 3x$$

4 Find the first 3 terms, in ascending powers of x, of the binomial expansion of $(3 - x)^6$ and simplify each term. **(4 marks)**

$$(3 - x)^6 = 3^6 + \binom{6}{1} \times 3^5 \times -x$$
$$+ \binom{6}{2} \times 3^4 \times -x^2 \ldots$$
$$= 729 - 1458x - 1215x^2 \ldots$$

5 Given that $0 < x < 4$ and
$$\log_5(4 - x) - 2\log_5 x = 1,$$
find the value of x. **(6 marks)**

$$\log_5\left(\dfrac{4 - x}{2x}\right) = 1$$

$$\dfrac{4 - x}{2x} = 5$$

$$4 - x = 10x$$

$$11x = 4$$

$$x = \dfrac{4}{11}$$

Checking your work

If you have time left in your exam you should check back through your working:

- ☑ Check you have answered **every question part**.
- ☑ Make sure your answers are given in the **correct form**.
- ☑ Double-check numerical calculations such as **binomial coefficients**.
- ☑ Cross out any incorrect working with a **single neat line** and underline the correct answer.

Now try this

Find the mistake in each student answer on this page, and write out the correct working for each question. Turn over for the answers.

You are still the examiner!

Before looking at this page, turn back to page 54 and try to spot the key mistake in each student's working. Use this page to check your answers. The corrections are shown in red and these answers are now 100% correct.

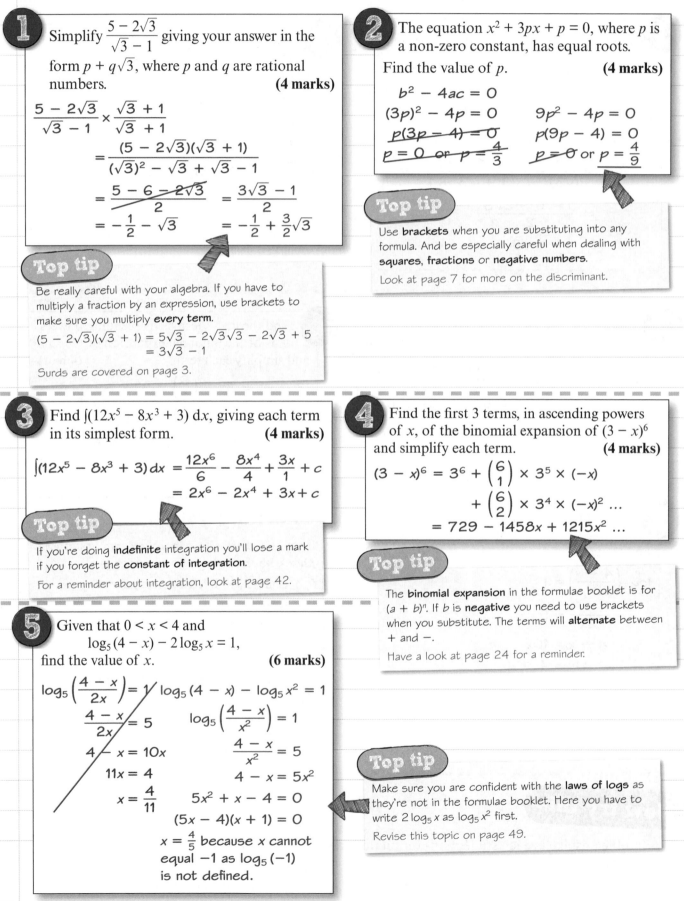

1 Simplify $\dfrac{5 - 2\sqrt{3}}{\sqrt{3} - 1}$ giving your answer in the form $p + q\sqrt{3}$, where p and q are rational numbers. **(4 marks)**

$$\dfrac{5 - 2\sqrt{3}}{\sqrt{3} - 1} \times \dfrac{\sqrt{3} + 1}{\sqrt{3} + 1}$$

$$= \dfrac{(5 - 2\sqrt{3})(\sqrt{3} + 1)}{(\sqrt{3})^2 - \sqrt{3} + \sqrt{3} - 1}$$

$$= \dfrac{5 - 6 - 2\sqrt{3}}{2} \qquad = \dfrac{3\sqrt{3} - 1}{2}$$

$$= -\dfrac{1}{2} - \sqrt{3} \qquad = -\dfrac{1}{2} + \dfrac{3}{2}\sqrt{3}$$

Top tip

Be really careful with your algebra. If you have to multiply a fraction by an expression, use brackets to make sure you multiply **every term**.

$(5 - 2\sqrt{3})(\sqrt{3} + 1) = 5\sqrt{3} - 2\sqrt{3}\sqrt{3} - 2\sqrt{3} + 5$
$= 3\sqrt{3} - 1$

Surds are covered on page 3.

2 The equation $x^2 + 3px + p = 0$, where p is a non-zero constant, has equal roots. Find the value of p. **(4 marks)**

$$b^2 - 4ac = 0$$
$$(3p)^2 - 4p = 0 \qquad 9p^2 - 4p = 0$$
$$p(3p - 4) = 0 \qquad p(9p - 4) = 0$$
$$p = 0 \text{ or } p = \dfrac{4}{3} \qquad p = 0 \text{ or } p = \dfrac{4}{9}$$

Top tip

Use **brackets** when you are substituting into any formula. And be especially careful when dealing with **squares**, **fractions** or **negative numbers**.

Look at page 7 for more on the discriminant.

3 Find $\int(12x^5 - 8x^3 + 3)\,dx$, giving each term in its simplest form. **(4 marks)**

$$\int(12x^5 - 8x^3 + 3)\,dx = \dfrac{12x^6}{6} - \dfrac{8x^4}{4} + \dfrac{3x}{1} + c$$

$$= 2x^6 - 2x^4 + 3x + c$$

Top tip

If you're doing **indefinite** integration you'll lose a mark if you forget the **constant of integration**.

For a reminder about integration, look at page 42.

4 Find the first 3 terms, in ascending powers of x, of the binomial expansion of $(3 - x)^6$ and simplify each term. **(4 marks)**

$$(3 - x)^6 = 3^6 + \binom{6}{1} \times 3^5 \times (-x)$$

$$+ \binom{6}{2} \times 3^4 \times (-x)^2 \dots$$

$$= 729 - 1458x + 1215x^2 \dots$$

Top tip

The **binomial expansion** in the formulae booklet is for $(a + b)^n$. If b is **negative** you need to use brackets when you substitute. The terms will **alternate** between $+$ and $-$.

Have a look at page 24 for a reminder.

5 Given that $0 < x < 4$ and
$$\log_5(4 - x) - 2\log_5 x = 1,$$
find the value of x. **(6 marks)**

$$\log_5\left(\dfrac{4 - x}{2x}\right) = 1 \qquad \log_5(4 - x) - \log_5 x^2 = 1$$

$$\dfrac{4 - x}{2x} = 5 \qquad \log_5\left(\dfrac{4 - x}{x^2}\right) = 1$$

$$4 - x = 10x \qquad \dfrac{4 - x}{x^2} = 5$$

$$11x = 4 \qquad 4 - x = 5x^2$$

$$x = \dfrac{4}{11} \qquad 5x^2 + x - 4 = 0$$

$$(5x - 4)(x + 1) = 0$$

$$x = \dfrac{4}{5} \text{ because } x \text{ cannot}$$
equal -1 as $\log_5(-1)$ is not defined.

Top tip

Make sure you are confident with the **laws of logs** as they're not in the formulae booklet. Here you have to write $2\log_5 x$ as $\log_5 x^2$ first.

Revise this topic on page 49.

Sampling

In statistics, a **population** is a group of people you are interested in. A **sample** is a smaller group chosen from a larger population. In order to select a sample, you need to name or number the population to create a list called a **sampling frame**. Here are four types of sample you need to know about:

Random samples

 Simple random sample – pick names out of a hat or generate random numbers.

2 **Systematic sample** – number the population, and select sampling units at regular intervals. The first sampling unit is selected randomly.

Non-random samples

 Quota sample – select a predetermined number of sampling units with certain defined characteristics.

4 **Opportunity sample** – select most conveniently available sampling units that meet required criteria.

Worked example

A chicken farmer wants to test the shell thickness of a batch of 100 eggs.

Give one reason why a census might not be suitable for this test.   **(1 mark)**

The farmer might need to break the eggs to test them, and he would have none left to sell.

If the testing process destroys or consumes the sampling unit, then a census may not be appropriate.

Sample vs census

A **census** observes or measures every member of a population. Carrying out a census can be time-consuming and expensive, and can produce a large amount of data to clean and process.

A **sample** is quicker and cheaper than a census, but if the sample is not representative of the population then it can introduce bias.

Stratified sampling

If a population is split into groups, you can select a stratified sample by ensuring the number of elements selected from each group is proportional to the size of the group. To work out how many members to select from each group, find the **sampling fraction** and multiply this by the size of each group.

$$\text{Sampling fraction} = \frac{\text{Sample size}}{\text{Population size}}$$

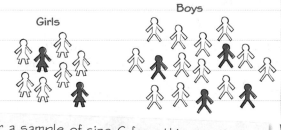

For a sample of size 6 from this population of size 24 you need to select $8 \times \frac{6}{24}$ girls and $16 \times \frac{6}{24} = 4$ boys.

Now try this

This table shows the number of employees in an engineering company. The CEO of the company wants to select a random sample of 50 employees, stratified by gender and office location.

	Bristol	London	Total
Male	119	91	210
Female	152	74	226
Total	271	165	436

(a) How many female employees from Bristol should she select for her sample?   **(2 marks)**

(b) Describe a suitable sampling frame.   **(1 mark)**

A stratified sample is an example of a random sample – all items have an equal chance of selection.

Mean

The mean is a measure of **central tendency**. The mean is sometimes written as $\bar{x}$. You might have to calculate an **estimate** of the mean of **grouped** data given in a **frequency table**.

Formulae for the mean

Learn these two formulae for the mean – they're not given in the formulae booklet.

 For n discrete data values, the mean is:

$$\bar{x} = \frac{\sum x}{n}$$
— The **sum** of the data values
— The **number** of data values

 For data given in a frequency table:

$$\bar{x} = \frac{\sum fx}{\sum f}$$
— The **sum** of (frequency × data value) or (frequency × midpoint)
— The **total** frequency

Grouped data

You can use formula 2 on the left with a **grouped frequency distribution**. The value of x for each group is the **midpoint** of that group.

Height, h (cm)	Midpoint
$0 \leqslant t < 5$	2.5
$5 \leqslant t < 7$	6
$7 \leqslant t < 10$	8.5

$\frac{0+5}{2} = 2.5$

$\frac{5+7}{2} = 6$

Watch out!

You can calculate the mean of a set of discrete data quickly using the statistics functions on your calculator. However, you need to understand the formula in case you are given summarised data.

Worked example

A researcher is studying the rainfall in Jacksonville. She uses the large data set to obtain the 24-hour rainfall total, x mm, each day in September 2015. She summarises her results as follows:

$$\sum x = 316.2$$

Calculate the mean 24-hour rainfall total in Jacksonville in September 2015. **(1 mark)**

$$\bar{x} = \frac{\sum x}{n} = \frac{316.2}{30} = 10.5 \text{ mm (3 s.f.)}$$

There are 30 days in September so $n = 30$.

There is more about the large data set on page 60.

Worked example

A teacher asked her class of 30 students how long, to the nearest hour, they spent on a homework project.

Hours	1–3	4–7	8–10	11–20
Frequency	7	12	8	3

Estimate the mean of the time spent on the project. **(2 marks)**

Midpoint	2	5.5	9	15.5
Frequency	7	12	8	3

$$\bar{x} = \frac{\sum fx}{\sum f}$$

$$= \frac{7 \times 2 + 12 \times 5.5 + 8 \times 9 + 3 \times 15.5}{30}$$

$$= \frac{198.5}{30} = 6.6 \text{ hours (1 d.p.)}$$

Start by working out the midpoint of each class interval. The midpoint of the 4–7 class is $\frac{3.5 + 7.5}{2} = 5.5$. In your exam you can estimate the mean quickly using your calculator.

There is more about using a calculator with grouped frequency tables on page 61.

Now try this

Jamie recorded the temperature at his school at midday, $x\,°C$, each day for 15 days. He summarised his results: $\sum x = 251$

Paul recorded the temperature in °C on the next 5 days.

Here are his results: 19   22   15   21   16

You know the sum of the first 15 values. Add the next 5 values then divide by the total number of values, 20.

(a) Calculate the mean temperature during the whole 20 days. **(2 marks)**

On the next day, the midday temperature was 16 °C.

(b) State, giving a reason, the effect this will have on the mean for the whole 21 days. **(1 mark)**

Median and quartiles

The **median** (Q_2), the **lower quartile** (Q_1), the **upper quartile** (Q_3) and the **percentiles** are all measures of **location**.

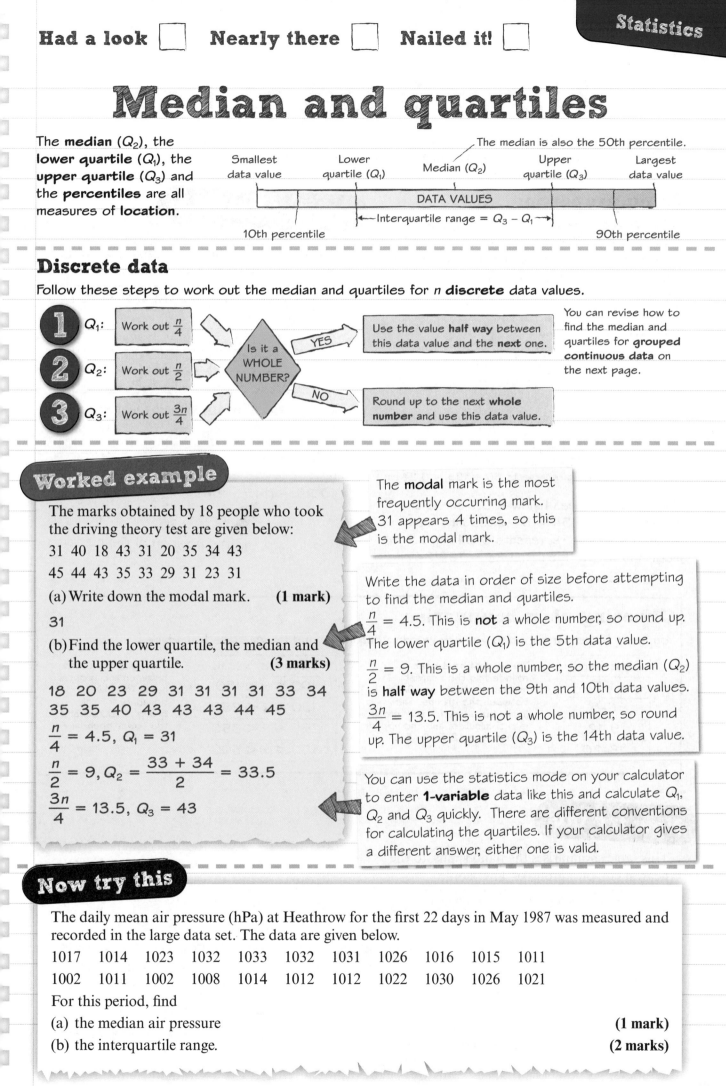

The median is also the 50th percentile.

| Smallest data value | Lower quartile (Q_1) | Median (Q_2) | Upper quartile (Q_3) | Largest data value |

DATA VALUES

Interquartile range = $Q_3 - Q_1$

10th percentile   90th percentile

Discrete data

Follow these steps to work out the median and quartiles for n **discrete** data values.

1. Q_1: Work out $\frac{n}{4}$
2. Q_2: Work out $\frac{n}{2}$
3. Q_3: Work out $\frac{3n}{4}$

Is it a WHOLE NUMBER?

YES → Use the value **half way** between this data value and the **next** one.

NO → Round up to the next **whole number** and use this data value.

You can revise how to find the median and quartiles for **grouped continuous data** on the next page.

Worked example

The marks obtained by 18 people who took the driving theory test are given below:

31 40 18 43 31 20 35 34 43
45 44 43 35 33 29 31 23 31

(a) Write down the modal mark. **(1 mark)**

31

(b) Find the lower quartile, the median and the upper quartile. **(3 marks)**

18 20 23 29 31 31 31 31 33 34
35 35 40 43 43 43 44 45

$\frac{n}{4} = 4.5$, $Q_1 = 31$

$\frac{n}{2} = 9$, $Q_2 = \frac{33 + 34}{2} = 33.5$

$\frac{3n}{4} = 13.5$, $Q_3 = 43$

The **modal** mark is the most frequently occurring mark. 31 appears 4 times, so this is the modal mark.

Write the data in order of size before attempting to find the median and quartiles.

$\frac{n}{4} = 4.5$. This is **not** a whole number, so round up. The lower quartile (Q_1) is the 5th data value.

$\frac{n}{2} = 9$. This is a whole number, so the median (Q_2) is **half way** between the 9th and 10th data values.

$\frac{3n}{4} = 13.5$. This is not a whole number, so round up. The upper quartile (Q_3) is the 14th data value.

You can use the statistics mode on your calculator to enter **1-variable** data like this and calculate Q_1, Q_2 and Q_3 quickly. There are different conventions for calculating the quartiles. If your calculator gives a different answer, either one is valid.

Now try this

The daily mean air pressure (hPa) at Heathrow for the first 22 days in May 1987 was measured and recorded in the large data set. The data are given below.

1017   1014   1023   1032   1033   1032   1031   1026   1016   1015   1011
1002   1011   1002   1008   1014   1012   1012   1022   1030   1026   1021

For this period, find

(a) the median air pressure   **(1 mark)**

(b) the interquartile range.   **(2 marks)**

Linear interpolation

You can use linear interpolation to estimate the **median**, **quartiles** and **percentiles** of grouped continuous data. You don't know the exact data values, so you assume the data values are **evenly distributed** within each group.

Using proportion

In the Worked example on the right, the median is in the 101–130 cm group. To use interpolation, you assume that the heights in this group are **evenly distributed** between the **lower** and **upper class boundaries**.

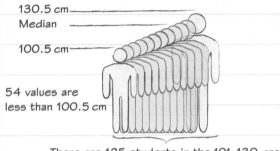

130.5 cm
Median
100.5 cm

54 values are less than 100.5 cm

There are 125 students in the 101–130 group, so each student represents 30 ÷ 125 = 0.24 cm

There are 54 values below 100.5 cm so the median is 145 − 54 = 91 values into the group. Each value represents 0.24 cm, so the median is 100.5 + 91 × 0.24 = 122.34 cm.

This table shows the heights, to the nearest cm, of 290 students in a school.

Height (cm)	Number of students
71–100	54
101–130	125
131–160	87
161–190	24

Use interpolation to estimate the median height. **(2 marks)**

$$\frac{290}{2} = 145$$

$$100.5 + (145 - 54) \times \frac{30}{125} = 122.34 \text{ cm}$$

Work out $\frac{n}{2}$ to find the position of the median.

Choosing class boundaries

You need to be careful when you are working out lower and upper class boundaries.

Time (mins)
1–5
6–8
9–12
13–20

These data have been rounded to the nearest minute and there are GAPS between the groups. The lower boundary for this group is 8.5 minutes, and the upper boundary is 12.5 minutes.

Distance, x (cm)
$100 \leqslant x < 150$
$150 \leqslant x < 250$
$250 \leqslant x < 350$
$350 \leqslant x < 500$

There are NO GAPS between the groups here. The lower boundary for this group is 100 cm and the upper boundary is 150 cm.

This table summarises the hand spans, to the nearest cm, of a group of 178 people.

Hand span (cm)	Frequency
10–14	29
15–17	64
18–19	55
20–22	21
23–30	9

Use interpolation to estimate the median Q_2, the lower quartile Q_1 and the 90th percentile of these data. **(4 marks)**

You can use interpolation to estimate the quartiles in exactly the same way as you estimate the median.

$\frac{n}{4} = 44.5$, so Q_1 is 44.5 − 29 = 15.5 values into the 15–17 group. This group is 17.5 − 14.5 = 3 cm wide, so each value represents $\frac{3}{64}$ cm. The estimate for Q_1 is

$$14.5 + (44.5 - 29) \times \frac{3}{64}$$

Round your answers to 2 decimal places.

Standard deviation 1

Standard deviation and **variance** are both measures of **spread** (or **dispersion**). Standard deviation is used more frequently because it has the **same units** as the data.

Discrete data

For n discrete data values:

$$\text{Variance} = \frac{\sum x^2}{n} - \left(\frac{\sum x}{n}\right)^2$$

Standard deviation = SD = $\sqrt{\text{Variance}}$

You can revise how to find the variance and standard deviation for **grouped continuous data** on the next page.

Notation

You need to be familiar with the notations used in statistics formulae.

- ☑ $\sum x^2$ means the sum of the squares of each of the data values.
- ☑ σ is sometimes used for the standard deviation.
- ☑ σ^2 is sometimes used for the variance.

Worked example

The ages, x years, of 50 passengers on a coach were recorded.

Given that $\sum x = 1756$ and $\sum x^2 = 69\,942$, calculate the standard deviation of the ages. **(2 marks)**

$$\text{Variance} = \frac{69942}{50} - \left(\frac{1756}{50}\right)^2 = 165.4256$$

$$\text{SD} = \sqrt{165.4256} = 12.9 \text{ years (3 s.f.)}$$

Summary statistics

You will often be given summary statistics like $\sum x$, $\sum x^2$ and S_{xx} in your exam. You can use these to simplify your calculations:

$$S_{xx} = \sum x^2 - \frac{(\sum x)^2}{n}, \text{ so another}$$

formula for the **variance** is $\frac{S_{xx}}{n}$.

The large data set

Some of the questions in your statistics exam will be based on meteorological data from the large data set. You don't need to learn any data from the large data set, but you will need to be **familiar** with it. This means:

- ☑ Knowing the approximate locations of each of the eight weather stations
- ☑ Understanding the different categories of data and the approximate values they can take
- ☑ Recognising coastal and non-coastal locations
- ☑ Knowing that high humidity can give rise to misty and foggy conditions
- ☑ Identifying likely locations based on given statistics (such as knowing that it is generally warmer in Beijing than in UK locations)

Now try this

The daily maximum temperature, $t\,°C$, in Camborne for the month of May 1987 was recorded and the data were summarised as follows:

$$\sum t = 397.3 \text{ and } \sum t^2 = 5165.39$$

(a) Calculate the mean and standard deviation, giving both answers correct to 1 d.p. **(2 marks)**

The mean and standard deviation of the daily maximum temperature in Leuchars for the same period were 13.6 °C and 1.2 °C respectively.

(b) State, with a reason, whether these results support the conclusion that southern locations are warmer than northern locations. **(2 marks)**

Standard deviation 2

You can **estimate** the variance and standard deviation of **grouped continuous data** given in a frequency table using the **midpoints**, x, of each group.

Grouped continuous data

For data given in a frequency table:

$$\text{Variance} = \frac{\sum fx^2}{\sum f} - \left(\frac{\sum fx}{\sum f}\right)^2 = \frac{\sum fx^2}{n} - (\bar{x})^2$$

Standard deviation = SD = $\sqrt{\text{Variance}}$

Learn these formulae – they are **not** in the booklet.

Notation

$\sum fx^2$ is the **sum** of (frequency × midpoint²). This is **not** the same as $\left(\sum fx\right)^2$.

For a reminder about the other notation used in the formula for variance, have a look at pages 57 and 60.

Worked example

This table shows the times a random sample of 150 people took to complete a puzzle.

Time, t (mins)	Number of people	Midpoint
$0 \leqslant t < 5$	23	2.5
$5 \leqslant t < 7$	82	6
$7 \leqslant t < 10$	36	8.5
$10 \leqslant t < 15$	9	12.5

Use your calculator to estimate the mean and the standard deviation of the length of time taken. **(3 marks)**

$\bar{t} = 3.34$ mins (3 s.f.)

$\sigma = 2.39$ mins (3 s.f.)

Use the midpoint of each class interval as your x-value for that class. You could work out the standard deviation using the formulae given above:

$\sum f = 290$

$\sum fx = 23 \times 2.5 + 82 \times 6 + 36 \times 8.5 + 9 \times 12.5 = 968$

$\sum fx^2 = 23 \times 2.5^2 + 82 \times 6^2 + 36 \times 8.5^2 + 9 \times 12.5^2 = 7103$

$\sigma = \sqrt{\dfrac{7103}{150} - \left(\dfrac{968}{150}\right)^2} = 2.39$ (3 s.f.)

However, it is quicker to work it out using the statistical functions on your calculator.

Calculator skills

You need to be confident using the statistical functions on your calculator. If you have to enter a grouped frequency distribution like the one above into your calculator, you should enter the **midpoints** of each class interval as the x-values. You might need to make sure you have switched on **frequency mode** on your calculator to enter a frequency distribution.

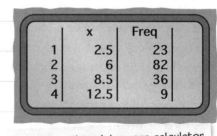

Enter data values into your calculator very carefully!

Now try this

A researcher timed how long, to the nearest minute, a group of shoppers spent in a supermarket checkout queue.

Minutes	1–3	4–6	7–10	11–20
Frequency	48	31	15	6

Use your calculator to estimate the mean and standard deviation. **(3 marks)**

Coding

Coding is a way of making statistics calculations easier. Each data value is **coded** to make a new set of data values. It is especially useful when you are dealing with **very large** or **very small** data values.

Transforming data

You can use this formula to transform a data value x into a new data value y:

$$y = \frac{x - a}{b}$$

The numbers a and b are chosen to make the final y-values easy to work with.

Effects of coding

If data is coded, different statistics will change in different ways.

	Before	After
Data values	x	$\frac{x - a}{b}$
Mean	$\bar{x}$	$\frac{\bar{x} - a}{b}$
Median	m	$\frac{m - a}{b}$
Standard deviation	s	$\frac{s}{b}$

Worked example

A scientist measures the temperature, x, in a nuclear reactor.

She uses the coding $y = \dfrac{x - 300}{10}$ to code her data.

The mean of the coded data is $\bar{y} = 3$ and the standard deviation of the coded data is $\sigma_y = 1.72$.

Calculate the mean and standard deviation of the original data. **(2 marks)**

$3 = \dfrac{\bar{x} - 300}{10}$ so $\bar{x} = 30 + 300$

$\qquad\qquad\qquad\qquad = 330\,°C$

$1.72 = \dfrac{\sigma_x}{10}$ so $\sigma_x = 17.2\,°C$ (3 s.f.)

Problem solved!

You will often have to calculate the mean or standard deviation of the **original** data, based on information about the **coded** data. You need to rearrange the rules given above:

$$\bar{y} = \frac{\bar{x} - a}{b} \text{ so } \bar{x} = b\bar{y} + a$$

$$\sigma_y = \frac{\sigma_x}{b} \text{ so } \sigma_x = n\sigma_y$$

You will need to use problem-solving skills throughout your exam – **be prepared!**

Watch out! The **standard deviation** measures **spread** so it is only affected when the data are multiplied or divided by a constant. Adding or subtracting a constant has no effect.

Now try this

1 From the large data set, the daily mean visibility, x metres, in Leeming between May and October 1987 was recorded.

The data is coded using the coding $y = \dfrac{x}{100}$ and the following summary statistics are obtained:

$n = 184$, $\sum y = 4380$, $S_{yy} = 2727.3$

Find the mean and standard deviation of the daily mean visibility in Leeming between May and October 1987. **(4 marks)**

Write down a coding formula that corresponds to a reduction of 8 followed by a 10% increase.

2 A group of 10 students took a computer-marked maths test. Their scores, x, are summarised as follows:

$\sum x = 668 \qquad \sum x^2 = 47\,870$

(a) Find the mean and standard deviation of the marks. **(3 marks)**

An instructor discovered that a computer error had caused one question to be marked correct on all the students' tests. He reduced each student's mark by 8, and then increased it by 10%.

(b) Find the mean and standard deviation of the adjusted marks. **(2 marks)**

Box plots and outliers

A box plot shows the **maximum** and **minimum** values and the **quartiles** of a distribution. It is usually drawn on graph paper with a scale.

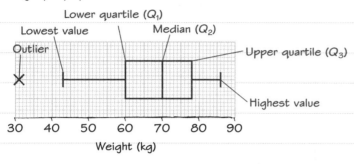

Weight (kg)

Half the weights were between 60 kg and 78 kg. 25% of the weights were less than 60 kg.

Outliers

A value which doesn't fall within the main body of the data is called an **outlier**. One common definition of an outlier is:

✓ values less than $Q_1 - 1.5 \times IQR$

✓ values greater than $Q_3 + 1.5 \times IQR$

On the diagram on the left, the IQR is $78 - 60 = 18$. The data value 31 is an outlier because it is less than $60 - 1.5 \times 18 = 33$.

Outliers are sometimes removed, or **cleaned**, from a data set. There is more about this on page 67.

Worked example

Here is a table showing information about the test scores for all the students in a school.

Two lowest values	2, 5
Lower quartile	10
Median	15
Upper quartile	26
Two highest values	40, 47

An outlier is a value that is greater than Q_3 plus 1.0 times the interquartile range or less than Q_1 minus 1.0 times the interquartile range.

Draw a box plot to represent these data, indicating clearly any outliers. **(5 marks)**

$IQR = 26 - 10 = 16$

$Q_1 - 1.0 \times IQR = -6$ $Q_3 + 1.0 \times IQR = 42$

Mark

Read the question carefully – any rule used to specify outliers will **always** be given in the question. You need to work out the lower and upper boundaries for the outliers before deciding which data values (if any) are outliers. 47 is greater than $Q_3 + 1.0 \times IQR$ so using this definition it is an outlier – you need to mark it on your box plot with a cross.

Whiskers and outliers

When there is an outlier, you can use either of these as your 'highest' or 'lowest' value:

1 the highest or lowest data value which is **NOT** an outlier

This is shown in the Worked example on the left.

2 the boundary of the normal data values that are not outliers.

In the Worked example, you could also draw the right-hand whisker up to 42.

Now try this

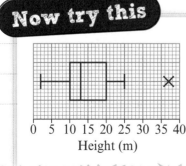

Height (m)

The box plot shows a summary of the heights, in metres, of the trees in a park.

(a) (i) Write down the height that 75% of the trees are shorter than.

(ii) State the name given to this value. **(2 marks)**

In (b), write down the name, what it means **and** one way of determining it.

(b) Explain what you understand by the × on the box plot. **(2 marks)**

Had a look ☐　　Nearly there ☐　　Nailed it! ☐

Cumulative frequency diagrams

You can use a cumulative frequency diagram to find the median, quartiles and percentiles of grouped continuous data. Cumulative frequency diagrams are also a good way to **compare** continuous frequency distributions.

There is more about comparing distributions on page 66.

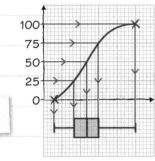

> You can use a cumulative frequency diagram to draw a box plot.

Cumulative frequency diagram of test results

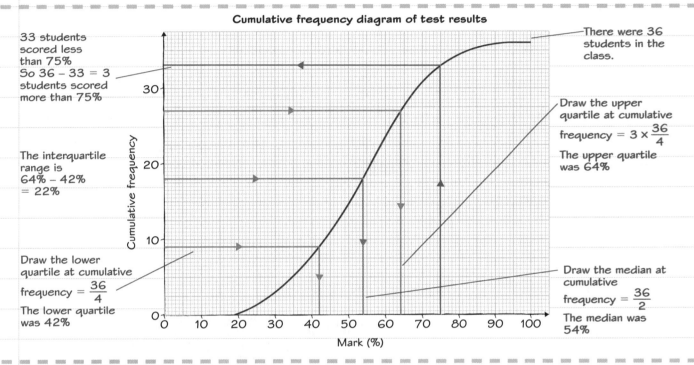

33 students scored less than 75%
So 36 – 33 = 3 students scored more than 75%

The interquartile range is
64% – 42%
= 22%

Draw the lower quartile at cumulative frequency = $\frac{36}{4}$
The lower quartile was 42%

There were 36 students in the class.

Draw the upper quartile at cumulative frequency = $3 \times \frac{36}{4}$
The upper quartile was 64%

Draw the median at cumulative frequency = $\frac{36}{2}$
The median was 54%

Now try this

This frequency table summarises the daily mean wind speeds in Perth between May and October 1987.

Wind speed, w (kn)	Frequency
$w < 2$	12
$2 \leqslant w < 4$	31
$4 \leqslant w < 6$	49
$6 \leqslant w < 8$	39
$8 \leqslant w < 10$	29
$10 \leqslant w < 12$	17
$w \geqslant 12$	7

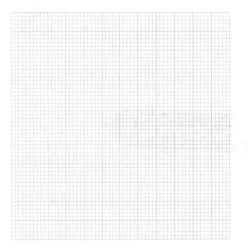

Given that the minimum wind speed recorded was 0.2 kn and the maximum was 21.8 kn,

(a) draw a cumulative frequency diagram to represent these data. **(3 marks)**

(b) Use your cumulative frequency diagram to estimate the 10th to 90th percentile range. **(3 marks)**

> To find the 10th percentile, read across from $\frac{n}{10}$ on the vertical axis to the curve, then down to the horizontal axis.

Histograms

Histograms are usually used to represent **grouped continuous data**. You probably won't be asked to draw a whole histogram in your exam, but you might have to **interpret** one, or make calculations about **widths** and **heights** of bars.

Worked example

The histogram shows the finishing times, to the nearest minute, of 40 runners in a 1500 m race.

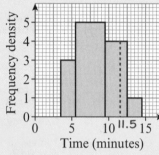

(a) Complete the table. **(2 marks)**

Finishing time (min)	Frequency
4–5	6
6–9	20
10–12	12
13–14	2

(b) Estimate the number of runners who finished the race in between 5.5 and 11.5 minutes. **(2 marks)**

$20 + \frac{2}{3} \times 12 = 28$

Your answer is only an estimate because you don't know how the data are distributed within each class.

Histogram facts

- ☑ No gaps between the bars.
- ☑ **area** of each bar is proportional to frequency.
- ☑ Vertical axis is labelled 'Frequency density'.
- ☑ Bars can be different widths.
- ☑ Frequency density $= \dfrac{\text{frequency}}{\text{class width}}$
- ☑ Bars are drawn between **class boundaries**.

Have a look at page 59 for a reminder about upper and lower class boundaries.

The 6–9 bar is plotted from 5.5 to 9.5, so it has width 4. Its frequency density is 5.

Frequency = frequency density × class width
$= 5 \times 4 = 20$

You can work out the frequency in the 10–12 class using the total number of runners $(40 - 20 - 6 - 2 = 12)$, or using the formula for frequency density.

Problem solved!

Be careful – 11.5 isn't one of the class boundaries. Frequency is proportional to **area** in a histogram. $\frac{2}{3}$ of the 10–12 bar is below 11.5, so you should include $\frac{2}{3}$ of these runners.

You will need to use problem-solving skills throughout your exam – **be prepared!**

Now try this

Alison weighed 50 apples, to the nearest gram. This table shows her results.

Weight (g)	55–57	58–63	64–68
Frequency	12	30	8

A histogram was drawn and the bar representing the 58–63 class was 4 cm wide and 6 cm high. For the 55–57 class, find

(a) the width

(b) the height

of the bar representing this class. **(3 marks)**

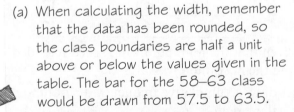

(a) When calculating the width, remember that the data has been rounded, so the class boundaries are half a unit above or below the values given in the table. The bar for the 58–63 class would be drawn from 57.5 to 63.5.

(b) Use the fact that the **area** of each bar is **proportional** to the frequency.

Had a look ☐   Nearly there ☐   Nailed it! ☐

Comparing distributions

If you need to compare two distributions, you can use **two** things.

1 Measures of **location** like the mean, median, mode and quartiles

Revise these on pages 57–59.

2 Measures of **spread**, like the range, interquartile range, variance, or standard deviation

Revise these on pages 58, 60 and 61.

Worked example

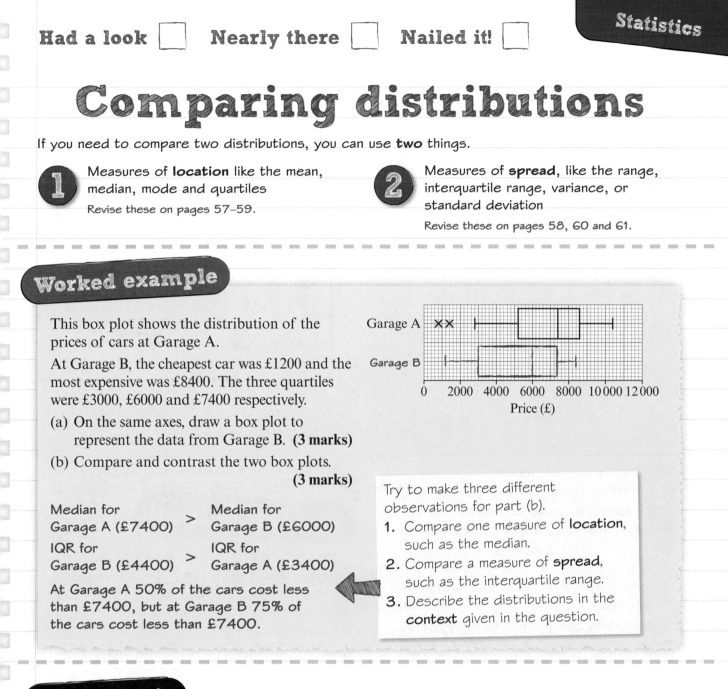

This box plot shows the distribution of the prices of cars at Garage A.

At Garage B, the cheapest car was £1200 and the most expensive was £8400. The three quartiles were £3000, £6000 and £7400 respectively.

(a) On the same axes, draw a box plot to represent the data from Garage B. **(3 marks)**

(b) Compare and contrast the two box plots. **(3 marks)**

Median for
Garage A (£7400) > Median for
Garage B (£6000)

IQR for
Garage B (£4400) > IQR for
Garage A (£3400)

At Garage A 50% of the cars cost less than £7400, but at Garage B 75% of the cars cost less than £7400.

Try to make three different observations for part (b).

1. Compare one measure of **location**, such as the median.

2. Compare a measure of **spread**, such as the interquartile range.

3. Describe the distributions in the **context** given in the question.

Now try this

The daily mean air temperature in °C is measured in Jacksonville in May 1987 and May 2015 and recorded in the large data set. The data are summarised as follows:

	Minimum	Q_1	Q_2	Q_3	Maximum	$\sum x$	$\sum x^2$
1987	19.5	22.2	23.2	23.9	25.8	710.9	16 364.77
2015	17.5	22.7	24.5	25.0	26.5	727.9	17 255.61

(a) Calculate the mean and standard deviation of the data for each of the two years. **(4 marks)**

(b) Compare the distributions for the two years. How far does the data support the conclusion that average air temperatures in Jacksonville have increased between 1987 and 2015? **(4 marks)**

The mean and standard deviation use all of the data values, whereas the median and IQR ignore extreme data values. If you are comparing data you should use either the mean together with the standard deviation, or the median together with the IQR. You could also consider the sample size when commenting on the conclusion.

Correlation and cleaning data

You can use correlation to describe a **linear relationship** between two variables. The closer the data approximates to a linear relationship, the **stronger** the correlation.

Worked example

The scatter diagram shows the maximum relative humidity, and mean visibility for the first 12 days in July, 2015 in Camborne.

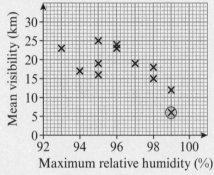

(a) Describe and interpret any correlation shown on the scatter diagram. **(2 marks)**

Quite strong negative correlation. As humidity increases, visibility decreases.

(b) From your knowledge of the large data set, explain whether the relationship between humidity and visibility might be causal. **(2 marks)**

High humidity can give rise to misty and foggy conditions, which can reduce visibility, so the relationship could be causal.

The visibility for the circled data point is found to be an outlier.

(c) Justify the inclusion of this outlier when analysing the data. **(1 mark)**

A visibility of 6 km is not unlikely, and follows the trend of the data.

Cause and effect

Watch out! Correlation doesn't always mean that two variables are related.

Bottled water doesn't cause bee stings but, when the weather is hotter, bottled water sales and bee stings both increase. This is an example of a non-causal relationship.

You need to understand what the different variables in the large data set represent, and how they might be related.

There is more on the large data set on page 60.

Problem solved!

Removing obviously incorrect data values is sometimes called **cleaning data**. You might want to **keep** an outlier if:
• it appears to be genuine
• the data is from a reliable source, or has already been cleaned.

You might want to **remove** an outlier if:
• it is obviously a mistake
• you want to intentionally omit extreme data values

You will need to use problem-solving skills throughout your exam – **be prepared!**

Now try this

Michelle is training a group of young people to use image-editing software. She records the time, y hours, it takes each person in the group to reach a certain standard of proficiency.

Group member	A	B	C	D	E	F	G	H	I	J
Age (x years)	15	19	81	16	20	17	19	21	18	15
Time (y hours)	14	13	10	12	7	18	9	6	9	11

Michelle determines that $x = 81$ is an outlier, and decides to omit the data for group member C from her results.

(a) Comment on the validity of Michelle's decision to omit the outlier. **(2 marks)**

(b) Draw a scatter diagram for the remaining 9 results. **(3 marks)**

(c) Describe the correlation shown on your scatter diagram, and interpret this in the context of the model. **(2 marks)**

Regression

A **regression line** on a scatter diagram is a type of line of best fit. It can be used as a **linear model** for the relationship between the two variables.

You don't need to be able to calculate regression lines in your exam, but you do need to be able to **interpret** their equations. The regression line (or **least squares regression line**) of y on x is written as:

$$y = a + bx$$

The coefficient b tells you the **gradient** of the regression line – this is how much the variable y increases for each unit increase in x.

Dependent and independent

The order of the variables in a regression equation is important.

In the regression model $y = a + bx$, x is the **independent** (or **explanatory**) variable, and y is the **dependent** (or **response**) variable. In an experiment, x would be the variable you changed, and y would be the variable you recorded. You always plot the explanatory variable on the **horizontal** axis.

If the value of b (the gradient) in the regression equation is **negative**, then the regression line slopes **downwards**. This means that as one variable increases, the other decreases.

An internet provider advertises an internet connection speed of 12 Mbps. Dhevan records the speed of 8 internet connections, s Mbps, at different distances, d km from the phone exchange. He records his results on a scatter diagram:

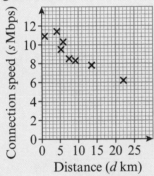

The equation of the regression line of s on d for these data is $s = 11.2 - 0.242d$

(a) Give an interpretation of the gradient of the regression line. **(1 mark)**

According to this model, for each additional km from the phone exchange, your connection speed will reduce by 0.242 Mbps.

(b) Explain why a regression model of the form $s = a + bd$ is supported for these data. **(1 mark)**

The data show strong negative correlation. Hence a linear model is suitable.

You could also consider whether the relationship is likely to be causal. Distance could affect connection speed, so this supports a relationship between the variables.

Validity

The **strength of the correlation** between two variables tells you how accurately a **linear** model represents the relationship between them. The stronger the correlation, the more closely the regression line models the data.

The scatter diagram shows the daily mean visibility, v km, and daily mean air pressure p hPa, in Hurn for the first 15 days of August 2015.

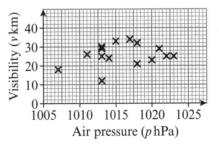

The equation of the regression line of v on p for these data is $v = -353.5 + 0.373p$

(a) Interpret the value 0.373 in this model. **(1 mark)**

(b) Comment on the validity of a linear regression model for these data. **(1 mark)**

Using regression lines

If you know the value of the **independent** variable, you can use a regression line (or its equation) to **estimate** or **predict** the corresponding value of the dependent variable. You won't need to make predictions like this in your exam, but you do need to know when they are **reliable**. This scatter diagram shows the mileage in miles per gallon (mpg) of 9 cars, and their engine capacities.

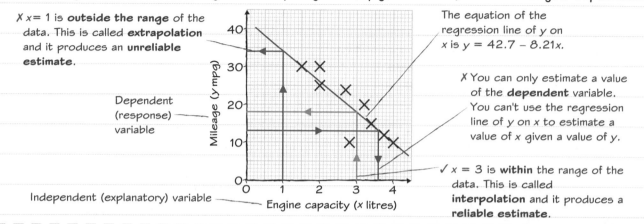

✗ x = 1 is **outside the range** of the data. This is called **extrapolation** and it produces an **unreliable estimate**.

Dependent (response) variable

Independent (explanatory) variable

The equation of the regression line of y on x is $y = 42.7 - 8.21x$.

✗ You can only estimate a value of the **dependent** variable. You can't use the regression line of y on x to estimate a value of x given a value of y.

✓ x = 3 is **within** the range of the data. This is called **interpolation** and it produces a **reliable estimate**.

Worked example

The age, x years, and shell size, y mm, of 8 Dungeness crabs were recorded.

y	151.8	150.4	140.3	133.4	155.9	153.3	141.6	127.7
x	3.3	3.0	2.4	2.3	3.3	3.0	2.7	2.2

The equation of the regression line of y on x for these data is $y = 83.17 + 22.03x$.

(a) Give an interpretation of the gradient of the regression line. **(1 mark)**

A typical crab shell will grow about 22 mm per year.

(b) Comment on the reliability of using this regression equation to estimate the shell size of a Dungeness crab with age:

(i) 1.6 years   (ii) 2.5 years   **(2 marks)**

(i) 1.6 years is outside the range of the data, so this is an example of extrapolation. The estimate will be unreliable.

(ii) 2.5 years is inside the range of the data, so this is an example of interpolation. The estimate will be reliable.

(d) Give a reason why this regression model would not be suitable for estimating the age of a Dungeness crab with a shell size of 145 mm. **(1 mark)**

Age is the independent variable in this model. You should only use a regression model to estimate values of the dependent variable.

Now try this

In a study, participants exercise for 5 minutes then record their breath rate (r breaths per minute) and pulse (p beats per minute). The data for 10 participants are shown in the scatter diagram.

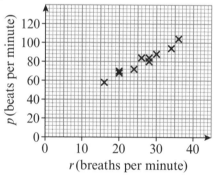

The equation of the regression line of p on r for these data is $p = 25.5 + 2.09r$

(a) Give a reason why a linear regression model is likely to be suitable for these data. **(1 mark)**

(b) The equation of the regression line is used to estimate the pulse rate of a participant making

(i) 22 breaths per minute

(ii) 10 breaths per minute.

Comment on the reliability of these estimates. **(2 marks)**

You would need to use the regression line of **x on y** to estimate age given shell size.

69

Drawing Venn diagrams

You can use a Venn diagram to represent different **events** in a **sample space**.

This Venn diagram shows the results when 50 people were surveyed about whether they owned a dog (*D*) or a cat (*C*). The rectangle represents the whole sample space (*S*), and each event is represented by an oval.

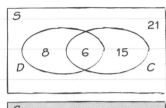

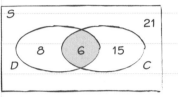

6 people owned a dog **and** a cat. You can write this event as '*D* and *C*'.

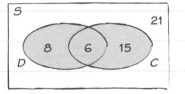

8 + 6 + 15 = 29 people owned a dog **or** a cat. You can write this event as '*D* or *C*'.

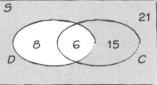

15 + 21 = 36 people **did not** own a dog. You can write this event as '**not** *D*'.

Worked example

200 students were surveyed about whether they are taking history, German or physics A-Level:

- 87 take history
- 30 take German
- 49 take physics
- 12 take history and German
- 26 take history and physics
- 8 take German and physics
- 2 take all three subjects.

Draw a Venn diagram to represent this information. **(5 marks)**

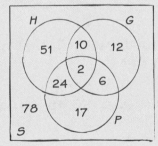

H and C and P
You can fill in the centre of the Venn diagram first. 2 students take all three subjects.

H and C but not P
The 12 students who take history and German **include** the 2 students who take all three subjects. So 12 − 2 = 10 students take history and German but **not** physics.

H but not C and not P
The 87 students who take history **include** the 10 + 2 + 24 = 36 students who take at least one other subject. So 87 − 36 = 51 take **only** history.

None of H, C or P
51 + 12 + 17 + 10 + 24 + 6 + 2 = 122 students take at least one subject. So 200 − 122 = 78 take **none** of the three subjects.

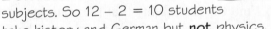

Now try this

An Indian restaurant recorded the bread choices at 100 tables. 56 tables ordered naan, 22 tables ordered roti and 40 ordered paratha. There were 8 tables that ordered naan and roti, 19 that ordered naan and paratha, and 15 that ordered roti and paratha. 6 tables ordered all three types of bread. Represent these data on a Venn diagram. **(5 marks)**

Make sure you label the whole sample space *S*, and draw three closed, intersecting circles or ovals to represent the three events. Label them *N*, *R* and *P*.

Using Venn diagrams

You can use Venn diagrams to help you answer **probability** questions. Find the region in the Venn diagram which **satisfies the conditions** in the question, add up the total number of **successful outcomes** and divide by the total number of possible outcomes.

Worked example

This Venn diagram shows the numbers of students in a class who play tennis, football and hockey.

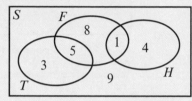

One of these students is selected at random.

(a) Show that the probability that the student plays more than one of the sports is $\frac{1}{5}$ **(2 marks)**

$3 + 5 + 8 + 9 + 1 + 4 = 30$

$$\frac{5 + 1}{30} = \frac{6}{30} = \frac{1}{5}$$

(b) Find the probability that the student plays either football or tennis, or both. **(2 marks)**

$$\frac{3 + 8 + 1 + 5}{30} = \frac{17}{30}$$

(c) Write down the probability that the student plays
  (i) tennis but not football
  (ii) all three sports. **(2 marks)**

(i) $\frac{3}{30} = \frac{1}{10}$ (ii) 0

Problem solved!

Make sure you add up the outcomes for **all** the regions you are interested in:

(a)   (b)   (c) (i)

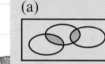

For part (c) (ii), there is no region where all three sports overlap, so no students played all three sports.

> You will need to use problem-solving skills throughout your exam – **be prepared!**

Worked example

For the events A and B, P(A and not B) = 0.27, P(B and not A) = 0.14 and P(A or B) = 0.72

(a) Draw a Venn diagram to illustrate the complete sample space for events A and B. **(3 marks)**

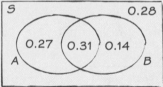

> The total of all the probabilities must add up to 1.

(b) Write down the value of P(A) and the value of P(B). **(3 marks)**

P(A) = 0.27 + 0.31 = 0.58
P(B) = 0.31 + 0.14 = 0.45

Now try this

A survey showed that 80% of students in a school own a laptop, and 22% of students own a tablet. 15% of students own neither a laptop nor a tablet.

(a) Draw a Venn diagram to represent this information. **(4 marks)**

A student is chosen at random.

(b) Write down the probability that this student owns a laptop but not a tablet. **(1 mark)**

> You can write probabilities as percentages on a Venn diagram. Remember that the percentages in the whole sample space need to add up to 100%.

Independent events

Two events are statistically independent if the outcome of one has **no effect** on the probability of the other.

Determining independence

Two events A and B are independent if and only if:

P(A and B) = P(A) × P(B)

This rule is given in the A-level section of the formulae booklet as:

$$P(A \cap B) = P(A) \, P(B)$$

where the symbol ∩ means 'and'.

You can use this rule in two ways:

 You can show that two events are independent by calculating P(A), P(B) and P(A and B) and demonstrating that the relationship is true.

 If you are told that two events are independent you can use the fact that P(A and B) = P(A) × P(B) to find unknown values.

> Remember to count **all** the possible outcomes in each event when determining the probability of that event. Make sure you don't use the rule given above in your working, as you haven't been told that the events are independent.

Worked example

This Venn diagram shows the numbers of students in a class of 30 who watched Coronation Street (C), Doctor Who (D) or EastEnders (E).

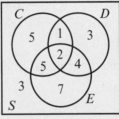

Determine whether watching EastEnders and watching Doctor Who are statistically independent. **(3 marks)**

$$P(D) = \frac{1 + 3 + 2 + 4}{30} = \frac{10}{30} = \frac{1}{3}$$

$$P(E) = \frac{5 + 2 + 4 + 7}{30} = \frac{18}{30} = \frac{3}{5}$$

$$P(D \text{ and } E) = \frac{2 + 4}{30} = \frac{6}{30} = \frac{1}{5}$$

$$P(D) \times P(E) = \frac{1}{3} \times \frac{3}{5} = \frac{1}{5} = P(D \text{ and } E)$$

So D and E are statistically independent.

Mutually exclusive events

Two events are mutually exclusive if they **cannot both** occur.
On the Venn diagram on the right, the events A and B are mutually exclusive because they **do not overlap**.

For two mutually exclusive events A and B: P(A and B) = 0 and P(A and B) = P(A) + P(B)

Now try this

The Venn diagram shows the probabilities of three events, A, B and C. x and y are unknown probabilities.

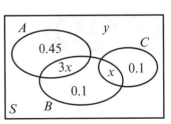

(a) Write down two events that are mutually exclusive. **(1 mark)**

Given that events B and C are independent,

(b) show that $400x^2 - 50x + 1 = 0$ **(3 marks)**

(c) find the values of x and y, justifying your choice of solution to the above quadratic equation. **(4 marks)**

Tree diagrams

Tree diagrams can be used to solve some probability problems. In your exam, it's usually only a good idea to draw a tree diagram when you are told to do so in the question.

Worked example

A computer virus is known to infect 7% of all computers. A software company writes a virus checker to determine whether or not a computer is infected. If a computer is infected, the test is positive with probability 0.95

If a computer is not infected, the test is positive with probability 0.1

(a) Draw a tree diagram to represent this information. **(3 marks)**

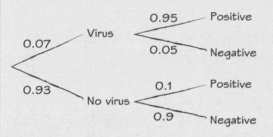

A computer is selected at random and tested.

(b) Find the probability that the test is positive. **(3 marks)**

P(Positive) = 0.07 × 0.95 + 0.93 × 0.1

= 0.1595

A sample of 10 computers is selected at random and tested.

(c) Find the probability that more than three of them test positive. **(3 marks)**

Let X = the number of computers that test positive. Then $X \sim B(10, 0.1595)$

$P(X > 3) = 1 - P(X \leqslant 3)$

= 1 − 0.9392...

= 0.06075...

So the probability is 0.0608 (3 s.f.)

Tree diagram checklist

Make sure that you:
- ☑ write a probability on **every** branch
- ☑ write an outcome at the **end** of every branch.

In your exam you **don't** need to:
- ☒ draw a tree diagram unless it's asked for in the question
- ☒ work out the probabilities of **all** the final outcomes – you will be asked for specific probabilities later in the question.

On each pair of branches the probabilities need to **add up to 1**.
So P(No virus) = 1 − 0.07 = 0.93

On a tree diagram you:

P(Virus and Positive) = 0.07 × 0.95

P(No virus and Positive) = 0.93 × 0.1

Add these together to find P(Positive).

Problem solved!

You are repeating trials with the same probability of success. You can model the number of computers that test positive with a **binomial distribution**. Write down the distribution and use your calculator to find the probability.

There is more about finding binomial probabilities on page 75

You will need to use problem-solving skills throughout your exam – **be prepared!**

Now try this

In a board game, Amy picks a question card. She can pick an easy, medium or hard question, with probabilities 0.54, 0.31 and 0.15 respectively.

The probabilities that she answers each type of question correctly are 0.8, 0.5 and 0.1 respectively.

(a) Draw a tree diagram to represent this information. **(3 marks)**

(b) Amy picks a card at random and answers the question. Work out the probability that she answers the question correctly. **(4 marks)**

Random variables

A **discrete random variable** can take a range of discrete **numerical values**. To define a random variable you need to know the range of values it can take (its **sample space**) and the probability that it takes each one. The probability that the random variable takes a certain value is given by its **probability function**.

The sum of probabilities

The most important fact you will use about random variables in your exam is that the probabilities of all the possible values of any random variable, X, always add up to 1. Another way of saying this is:

$$\sum_{\text{all } x}^{e} P(X = x) = 1$$

You always use **upper case** letters for random variables, and **lower case** letters for the values they can take. $P(X = x)$ means 'the probability that the random variable X takes the value x'.

Probability distributions

You can write the outcome from this spinner as a random variable X.

You can write its **probability distribution** in a table.

x	3	5	7
$P(X = x)$	$\frac{1}{6}$	$\frac{1}{3}$	$\frac{1}{2}$

This is the **sample space** for this random variable. X can only take these values.

$$\frac{1}{6} + \frac{1}{3} + \frac{1}{2} = 1$$

Its probability distribution could also be given using a **probability function**:

$$P(X = x) = \frac{x - 1}{12}, \quad x = 3, 5, 7$$

For example $P(X = 5) = \dfrac{5 - 1}{12} = \dfrac{4}{12} = \dfrac{1}{3}$

The random variable X has probability function

$$P(X = x) = \begin{cases} kx^2, & x = 1, 2, 3 \\ 2kx, & x = 4, 5 \end{cases}$$

where k is a constant.

(a) Find the value of k. **(2 marks)**

$k + 4k + 9k + 8k + 10k = 1$

$32k = 1$

$k = \dfrac{1}{32}$

(b) Find $P(X > 2)$. **(2 marks)**

$P(X > 2) = P(X = 3) + P(X = 4) + P(X = 5)$

$= \dfrac{9}{32} + \dfrac{8}{32} + \dfrac{10}{32} = \dfrac{27}{32}$

The curly bracket means that the probability function is different for different values of x. For $x = 1$, 2 and 3 you use $P(X = x) = kx^2$ and for $x = 4$ and 5 you use $P(X = x) = 2kx$.

For part (a), write an expression for the sum of the probabilities in terms of k. $\sum P(X = x) = 1$, so you can write an equation and solve it to find k.

For part (b), add up the probabilities for the values of X which make the inequality true.

For part (b), solve the inequality first.

1 The random variable Y has probability function

$$P(Y = y) = \frac{(y - 1)^2}{30}, \qquad y = 2, 3, 4, 5$$

(a) Construct a table giving the probability distribution of Y. **(3 marks)**

(b) Find $P(Y > 3)$. **(2 marks)**

2 The discrete random variable X has probability distribution given by

x	-2	-1	0	1	2
$P(X = x)$	0.1	a	0.15	$2a$	0.15

where a is a constant.

(a) Find the value of a. **(2 marks)**

(b) Find $P(3X + 1 \leqslant 6)$. **(2 marks)**

The binomial distribution

If you are carrying out a large number of trials you can model the number of **successful trials**, **X**, using a binomial distribution. For n trials, each with probability of success, p, you write:

$$X \sim B(n, p)$$

The probability that X takes a given value r is:

$$P(X = r) = \binom{n}{r} p^r (1 - p)^{n-r}$$

To bi or not to bi?

A binomial model is valid when

☑ there are a fixed number of trials

☑ the trials are independent

☑ there are two possible outcomes, with probabilities p and 1 − p

☑ the probability of each outcome is fixed.

Worked example

The discrete random variable $X \sim B(35, 0.82)$. Find:

(a) P(X = 29)

$$P(X = 29) = \binom{35}{29} 0.82^{29} 0.18^6$$
$$= 0.175 \ (3 \text{ s.f.})$$

(b) P(X ⩾ 25)

$$P(X \geqslant 25) = 1 - P(X \leqslant 24)$$
$$= 1 - 0.03877\ldots$$
$$= 0.961 \ (3 \text{ s.f.})$$

The easiest way to find binomial probabilities is using the binomial functions on your calculator. To find

```
Binomial PD
  x    :29
  N    :35
  p    :0.82
```

the probability that X takes a **single value** use the "Binomial probability distribution" function. You can also use the formula for P(X = r) and the nCr function on your calculator to find a single binomial probability.

To find the probability that X is **less than or equal to** a given value, use the "Binomial cumulative distribution" function. The question asks for P(X ⩾ 25), so use the fact that the sum of the probabilities is equal to 1. To find P(X ⩽ 24) type in:

```
Binomial CD
  x    :24
  N    :35
  p    :0.82
```

X can only take whole numbered values, so P(X < 4) = P(X ⩽ 3).

Worked example

A chicken farmer claims that 7% of his eggs have double yolks. The farmer takes a random sample of 20 eggs.

(a) Find the probability that fewer than 4 of them have double yolks.   **(2 marks)**

X = the number of double yolks in sample

$$X \sim B(20, 0.07)$$

$$P(X \leqslant 3) = 0.953 \ (3 \text{ s.f.})$$

(b) Give a reason why it would not be appropriate for the farmer to test her claim using a census.   **(1 mark)**

If she tested all the eggs she would have none left to sell.

Now try this

1 A fair six-sided dice is rolled 50 times. The discrete random variable X represents the number of 1s rolled.

(a) Justify the use of a binomial distribution to model X.   **(2 marks)**

(b) Find (i) P(X = 10) (ii) P(X < 7)   **(3 marks)**

2 A bag contains 60 blue counters and 40 red counters. Emma selects 10 counters at random from the bag without replacement. She models the number of blue counters selected as B(10, 0.6). Explain why this is not a suitable model for this situation.   **(2 marks)**

Hypothesis testing

You need to be able to carry out a hypothesis test for the probability, p, in a binomial distribution. Follow these steps to carry out a hypothesis test.

| Model the test statistic and define null (H_0) and alternative (H_1) hypotheses. | ⟹ | Assume H_0 is true and calculate the probability of the observed value (or a greater / lesser value) occurring | ⟹ | Compare this probability with a given significance level and write a conclusion stating whether H_0 is accepted or rejected. |

Worked example

A microchip manufacturer knows that 9% of the microchips produced using a certain process contain defects. The manufacturer trials a new manufacturing process. A sample of 50 chips from the new process are selected and 2 of them are observed to be faulty.

Test, at the 10% significance level, whether there is evidence that the proportion of faulty chips has reduced under the new process. State your hypotheses clearly. **(6 marks)**

Let X = the number of faulty chips in a sample of 50. Then $X \sim B(50, p)$.

$H_0: p = 0.09$, $H_1: p < 0.09$

Assume H_0 is true, so $X \sim B(50, 0.09)$.

$P(X \leq 2) = 0.1605...$

16% > 10% so there is not enough evidence to reject H_0.

The proportion of faulty chips has not significantly reduced under the new process.

How many tails?

✓ If you want to test whether p is likely to be **greater than** or **less than** a particular value you need to use a one-tailed test. For example:

$$H_0: p = 0.4, \ H_1: p > 0.4$$

✓ If you want to test whether p is likely to be **different** from a particular value, you need to use a two-tailed test. For example:

$$H_0: p = 0.75, \ H_1: p \neq 0.75$$

Problem solved!

You want to test whether the proportion has **reduced** so this is a **one-tailed test**.

You can use your calculator to find $P(X \leq 2)$ directly. Since the probability of this observation (or worse) is **greater than** 10% you **do not** reject H_0.

 You will need to use problem-solving skills throughout your exam – **be prepared!**

Now try this

1 A random variable has distribution $X \sim B(28, p)$. A single observation of $x = 25$ is taken from this distribution. Test, at the 2% significance level, $H_0: p = 0.7$ against $H_1: p > 0.7$. **(5 marks)**

This is a **two-tailed test** so you need to **halve** the significance level at each end of the test. The national pass rate is 49.3% so you would expect about 22 students to pass. Calculate $P(X \geq 29)$ and compare the result with 2.5%.

2 Between April 2015 and March 2016, the national average pass rate for the car driving theory exam was 49.3%. From the same period, a sample of 45 students in Edinburgh was taken, and 29 of them were found to have passed the exam.

Test, at the 5% significance level, whether the average pass rate in Edinburgh was different from the national average pass rate. You must state your hypotheses clearly. **(6 marks)**

Critical regions

In a hypothesis test, you can find the observed values of a random variable which would cause you to reject the null hypothesis. This set of values is called the **critical region**.

 One-tailed tests

For $X \sim B(15, p)$, if you are testing $H_0: p = 0.6$ against $H_1: p < 0.6$ at the 10% level, the critical region is $X \leqslant 6$.

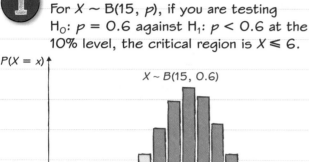

$P(X = x)$

$X \sim B(15, 0.6)$

0 1 2 3 4 5 6 7 8 9 10 11 12 13 14 15 x

$P(X \leqslant 6) = 0.09505$, which is the first value of x with $P(X \leqslant x) < 0.1$. The value 6 is called the **critical value**.

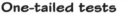 **Two-tailed tests**

For $X \sim B(12, p)$, if you are testing $H_0: p = 0.35$ against $H_1: p \neq 0.35$ at the 10% level, the critical region is $X \leqslant 1$ and $X \geqslant 8$.

$P(X = x)$

$X \sim B(12, 0.35)$

0 1 2 3 4 5 6 7 8 9 10 11 12 x

Divide the significance level by 2. The probability in **each** tail should be $\leqslant 0.05$. $P(X \leqslant 1) = 0.04244$ and $P(X \geqslant 8) = 0.02551$. There are two **critical values**: 1 and 8.

Worked example

The random variable X is modelled as $X \sim B(20, p)$. A single observation of X is taken and used to test $H_0: p = 0.3$ against $H_1: p > 0.3$.

(a) Using a 5% significance level, find the critical region for this test. **(2 marks)**

Assume H_0 is true, so $X \sim B(20, 0.3)$

$P(X \geqslant 9) = 1 - P(X \geqslant 8) = 1 - 0.8867$
$= 0.1133$

$P(X \geqslant 10) = 1 - P(X \leqslant 9) = 1 - 0.9520$
$= 0.0480$

The critical region is $X \geqslant 10$

(b) State the actual significance level of this test. **(1 mark)**

0.0480

The actual observed value of X is 11.

(c) Comment on this observation in light of your critical region. **(2 marks)**

11 lies inside the critical region, so this is evidence to reject H_0 at the 5% significance level.

Actual significance level

The actual probability that the observed value will fall within the critical region is sometimes called the actual significance level. This is also the probability that the null hypothesis is **rejected incorrectly**.

You are testing whether $p > 0.3$, so the critical value is the first value x such that $P(X \geqslant x) < 0.05$, or $P(X \geqslant x - 1) > 0.95$. You can use the **binomial cumulative distribution** table in the formulae booklet:

$p =$	0.25	0.30	0.35
$n = 20, x = 7$	0.8982	0.7723	0.6010
8	0.9591	0.8867	0.7624
9	0.9861	0.9520	0.8782
10	0.9961	0.9829	0.9468

Now try this

A single observation is taken from a binomial distribution $B(15, p)$, and is used to test $H_0: p = 0.45$ against $H_1: p \neq 0.45$

(a) Using a 2% level of significance, find the critical region for this test. The probability in each tail should be as close to 0.01 as possible. **(3 marks)**

(b) Find the actual significance level of this test. **(2 marks)**

Watch out! In part (a) you need to find probabilities **as close as possible** to 0.01. This means that the probabilities in each tail do not necessarily have to be less than 0.01.

Had a look ☐ Nearly there ☐ Nailed it! ☐

You are the examiner!

In your exam you might be asked to identify errors in working. You should also be confident **checking your own work**. Each of these students has made key mistakes in their working. Can you spot them all?

1 The Venn diagram shows how many of the 200 students at a performing arts school who dance (D), sing (G) and play a musical instrument (M).

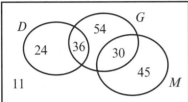

One of the students is selected at random.

(a) Find the probability that they either sing or play a musical instrument. **(2 marks)**

(b) Determine whether dancing and singing are statistically independent. **(3 marks)**

$11 + 24 + 36 + 54 + 30 + 45 = 200$

(a) P(G or M) = $\frac{30}{200}$ = 0.15

(b) P(D) = $\frac{24 + 36}{200}$ = 0.3

P(G) = $\frac{36 + 54}{200}$ = 0.45

P(G and D) = $\frac{36}{200}$ = 0.18

0.3 × 0.45 = 0.135 ≠ 0.18 so not independent

2 In a survey a random sample of 130 teenagers were asked how many hours, to the nearest hour, they spent online in the last week. The results are summarised in the table.

A histogram was drawn and the group (8–10) was represented by a rectangle that was 3 cm wide and 4 cm high.

Number of hours	Frequency
0–5	17
6–7	18
8–10	40
11–20	55

(a) Calculate the width and height of the rectangle representing the group (6–7). **(3 marks)**

(b) Use linear interpolation to estimate the median. **(4 marks)**

(a) 7.5 − 5.5 = 2

Frequency density = $\frac{18}{2}$ = 9

2 cm wide and 9 cm high.

(b) Median = 130 ÷ 2 = 65th value

17 + 18 = 35 data values in first two groups

8 + (65 − 35) × $\frac{2}{40}$ = 9.5

3 A clothing shop knows from experience that 35% of visitors will make a purchase. After redesigning the interior of the shop, the owner observes a random sample of 30 visitors and notes that 17 of them make a purchase.

Test, at the 2% significance level, whether there is evidence that the proportion of visitors who make a purchase has increased. **(6 marks)**

Let X = number of customers who make a purchase.

So $X \sim B(30, 0.35)$

P($X \geq 17$) = 1 − P($X \leq 17$)

= 0.0045

0.0045 < 0.02

So there is evidence that the proportion of visitors who make a purchase has increased.

Checking your work

If you have time left in your exam you should check back through your working:

✓ Check you have answered **every question part**.

✓ Double check any **binomial probabilities** using your calculator.

✓ Make sure any conclusions are given **in the context** of the question.

✓ Cross out any incorrect working with a **single neat line** and underline the correct answer.

Now try this

Find the mistakes in each student's answer, and write out the correct working for each question. Turn over for the answers.

You are still the examiner!

Before looking at this page, turn back to page 78 and try to spot the key mistakes in each student's working. Use this page to check your answers. The corrections are shown in red and these answers are now 100% correct.

1 The Venn diagram shows how many of the 200 students at a performing arts school dance (*D*), sing (*G*) and play a musical instrument (*M*).

One of the students is selected at random.

(a) Find the probability that they either sing or play a musical instrument. **(2 marks)**

(b) Determine whether dancing and singing are statistically independent. **(3 marks)**

(a) $P(G \text{ or } M) = \dfrac{36 + 54 + 30 + 45}{200}$

$= \cancel{0.15} \ 0.825$

(b) $P(D) = \dfrac{24 + 36}{200} = 0.3$

$P(G) = \dfrac{36 + 54 + 30}{200} = \cancel{0.45} \ 0.6$

$P(G \text{ and } D) = \dfrac{36}{200} = 0.18$

$\underset{0.6}{0.3} \times \underset{0.18}{\cancel{0.45}} = \cancel{0.27} \text{ so independent}$

Top tip

Make sure you read the question carefully and include **all** the possible outcomes in a given event. See pages 71 and 72.

2 In a survey a random sample of 130 teenagers were asked how many hours, to the nearest hour, they spent online in the last week. The results are summarised in the table.

A histogram was drawn and the group (8–10) was

Number of hours	Frequency
0–5	17
6–7	18
8–10	40
11–20	55

represented by a rectangle that was 3 cm wide and 4 cm high.

(a) Calculate the width and height of the rectangle representing the group (6–7). **(3 marks)**

(b) Use linear interpolation to estimate the median. **(4 marks)**

(a) $7.5 - 5.5 = 2$ Width = 2 cm

                       Height = 2.7 cm

$\dfrac{2h}{18} = \dfrac{3 \times 4}{40}$ so $h = 2.7$

(b) Median = $130 \div 2 = $ 65th value

17 + 18 = 35 data values in first two groups

$\cancel{8} + (65 - 35) \times \dfrac{\overset{3}{\cancel{2}}}{40} = \underset{9.5}{\overset{9.75}{\cancel{9.5}}}$

7.5

Top tip

Be really careful with class boundaries. The (8–10) group starts at 7.5, and has class width $10.5 - 7.5 = 3$
Revise this on page 59 and 65.

3 A clothing shop knows from experience that 35% of visitors will make a purchase. After redesigning the interior of the shop, the owner observes a random sample of 30 visitors and notes that 17 of them make a purchase.

Test, at the 2% significance level, whether there is evidence that the proportion of visitors who make a purchase has increased. **(6 marks)**

Let X = number of customers who make a purchase.

Then $X \sim B(30, p)$.

$H_0: p = 0.35$, $H_1: p > 0.35$

Assume H_0, so $X \sim B(30, 0.35)$

$P(X \geqslant 17) = 1 - P(X \leqslant \underset{16}{\cancel{17}})$

$= \cancel{0.0045} \ 0.0124$

$\underset{0.0124}{\cancel{0.0045}} < 0.02 \text{ so reject } H_0.$

So there is evidence that the proportion of visitors who make a purchase has increased.

Top tip

When carrying out a hypothesis test you need to state your hypotheses clearly, and state whether you have accepted or rejected H_0.
Revise hypothesis tests on page 76.
Binomial distributions are discrete, so $P(X \geqslant x) = 1 - P(X \leqslant x - 1)$.
There is more about this on page 75.

Modelling in mechanics

You can use a model to describe a problem mathematically. The model will make certain **assumptions** about the real-life situation to simplify the calculations. You might need to criticise or refine a model, and you should always interpret the model in the **context** given. This model shows the path of a basketball being thrown towards a hoop:

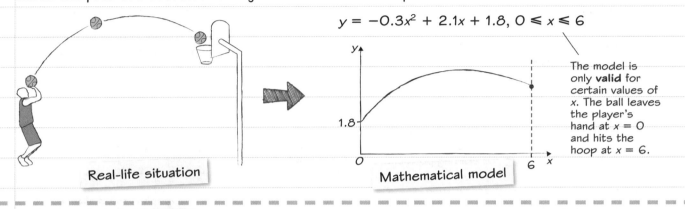

$$y = -0.3x^2 + 2.1x + 1.8, \; 0 \leqslant x \leqslant 6$$

The model is only **valid** for certain values of x. The ball leaves the player's hand at $x = 0$ and hits the hoop at $x = 6$.

Real-life situation   Mathematical model

Worked example

The diagram shows a car pulling a caravan. The car is attached to the caravan by a light, inextensible tow-bar. The caravan and car accelerate together at $2.5\,\text{ms}^{-2}$. State how you can use the assumption that the tow-bar is inextensible.   **(1 mark)**

The caravan and car will have the same acceleration.

An **inextensible** tow-bar, string or rod only affects **acceleration**. Don't talk about the tension in the tow-bar in your answer.

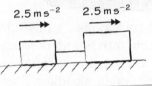

$2.5\,\text{m s}^{-2}$ $2.5\,\text{m s}^{-2}$

Pulleys and strings

Learn these three modelling facts for questions involving pulleys.

1 The pulley is **smooth** – the magnitude of the **tension** in the string will be the same on both sides of the pulley.

2 The string is **inextensible** – the magnitude of the **acceleration** will be the same for both particles.

3 The string is **light** – you can **ignore** the **weight** of the string in any calculations.

Motion under gravity

When an object moves freely under gravity, you ignore **air resistance**. This means the acceleration is **constant**. You also model objects as **particles**, so their weight acts at a single point. In your AS exam, you are also assuming that the motion is **vertical**.

Now try this

A block P of mass 0.8 kg is accelerating along a smooth horizontal table top. It is attached to a block Q of mass 0.1 kg by means of a light inextensible string running over a smooth pulley. P and Q are modelled as particles. State how you can use in your calculations the modelling assumptions that:

(a) the table top is smooth
(b) P and Q are modelled as particles
(c) the string is inextensible
(d) the string is light
(e) the pulley is smooth.   **(5 marks)**

Motion graphs

You need to understand and be able to sketch **displacement–time** graphs and **velocity–time** graphs for objects moving in straight lines.

Displacement–time graphs

The gradient of a displacement–time graph tells you the **velocity** of an object.

When the graph is horizontal the object is **stationary**.

Here the object is travelling with **constant** velocity.

The object has returned to its starting position.

A curve shows the object is accelerating or decelerating. Here the gradient is increasing so the speed is **increasing**.

Velocity–time graphs

This velocity–time graph shows motion of a train:

The train is **accelerating** here – the **gradient** of the graph is the acceleration.

If the graph is **flat** then the train is travelling at **constant** velocity.

The **area** under the graph is the **distance travelled**. The shaded area represents the distance travelled between 8 and 20 seconds.

Worked example

An athlete runs along a straight road. She starts from rest and moves with constant acceleration for 5 seconds, reaching a speed of $8\,\text{m s}^{-1}$. This speed is then maintained for T seconds. She then decelerates at a constant rate until she stops. She has run a total of $500\,\text{m}$ in $75\,\text{s}$.

Find the value of T.   **(3 marks)**

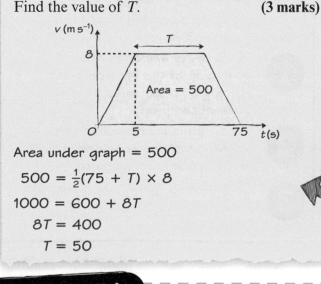

Area under graph = 500

$$500 = \tfrac{1}{2}(75 + T) \times 8$$

$$1000 = 600 + 8T$$

$$8T = 400$$

$$T = 50$$

Golden rules

Here are two important facts about velocity–time graphs.

1 Distance travelled = area under graph

In the example above, the train travels $\tfrac{1}{2}(10 + 16) \times 12 = 156\,\text{m}$ between $8\,\text{s}$ and $20\,\text{s}$.

2 Acceleration = gradient of graph

In the example above, between $8\,\text{s}$ and $20\,\text{s}$ the train is accelerating at $\frac{v}{t} = \frac{6}{12} = 0.5\,\text{m s}^{-2}$.

Problem solved!

You can solve this problem by sketching a velocity–time graph for the motion of the athlete. Use the formula for the area of a trapezium, $A = \tfrac{1}{2}(a + b)h$, to write an equation involving T.

You will need to use problem-solving skills throughout your exam – **be prepared!**

Now try this

A truck is moving along a straight horizontal road. At time $t = 0$, the truck passes a point A with speed $18\,\text{m s}^{-1}$. The truck moves with constant speed $18\,\text{m s}^{-1}$ until $t = 10\,\text{s}$. It then decelerates uniformly for 6 s. At time $t = 16\,\text{s}$, the speed of the truck is $V\,\text{m s}^{-1}$, and this speed is maintained until the truck reaches the point B at time $t = 30\,\text{s}$.

(a) Sketch a velocity–time graph to show the motion of the truck from A to B. **(3 marks)**

(b) Given that $AB = 455\,\text{m}$, find the value of V. **(5 marks)**

If the truck **decelerates**, the graph will slope down.

Constant acceleration 1

There are **five** formulae which describe the motion of a particle travelling in a **straight line** with constant acceleration. They are given in the formulae booklet but you should learn them. The other three formulae are covered on the next page.

Constant acceleration formulae

Here are the first two formulae you need to learn for motion in a straight line with constant acceleration.

1 $v = u + at$

2 $s = \frac{1}{2}(u + v)t$

Look at the diagram on the right to see what each letter represents.

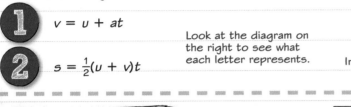

Initial velocity $u\,m\,s^{-1}$ Constant acceleration $a\,m\,s^{-2}$ Velocity after t seconds $v\,m\,s^{-1}$

Initial position   Displacement   Position after t seconds

s metres

Worked example

In taking off, an aircraft moves on a straight runway AB of length 1.2 km. The aircraft moves from A with initial speed $2\,m\,s^{-1}$. It moves with constant acceleration and 20 s later it leaves the runway at C with speed $74\,m\,s^{-1}$. Find

(a) the acceleration of the aircraft   **(2 marks)**

~~$s = ?$~~, $u = 2$, $v = 74$, $a = ?$, $t = 20$

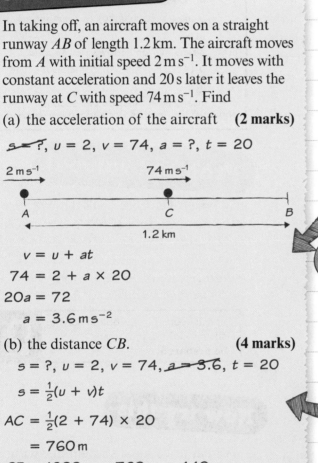

$v = u + at$

$74 = 2 + a \times 20$

$20a = 72$

$a = 3.6\,m\,s^{-2}$

(b) the distance CB.   **(4 marks)**

$s = ?$, $u = 2$, $v = 74$, ~~$a = 3.6$~~, $t = 20$

$s = \frac{1}{2}(u + v)t$

$AC = \frac{1}{2}(2 + 74) \times 20$

$= 760\,m$

$CB = 1200\,m - 760\,m = 440\,m$

Using *SUVAT*

The constant acceleration formulae are sometimes called the *SUVAT* formulae. In the exam you should write down all five letters.

✓ Write in any values you **know**.

✓ Put a **question mark** next to the value you want to find.

✓ **cross out** any values you don't need for that question.

This will help you choose which formula to use.

When you are using the **suvat** formulae **always** write down the formula **before** you substitute in.

Problem solved!

Read the question carefully. The distance AB is 1.2 km, but the aircraft does not take off at B. If you have to solve a constant acceleration question involving **three points** like this one, it's a good idea to draw a quick sketch to help you see what is going on. In part (b), s is the distance AC in metres. You need to subtract it from 1200 m to find the distance CB.

 You will need to use problem-solving skills throughout your exam – **be prepared!**

Now try this

A car moves along a straight stretch of road AB. The car moves with initial speed $2\,m\,s^{-1}$ at point A. It accelerates constantly for 12 seconds, reaching a speed of $23\,m\,s^{-1}$ at point B. Find

(a) the acceleration of the car   **(2 marks)**

(b) the distance AB.   **(2 marks)**

You can answer both parts of this question using the formulae given on this page. But you can use any of the other **suvat** formulae if you are confident with them.

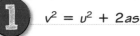

Constant acceleration 2

Here are three more formulae you can use to solve problems involving constant acceleration in a **straight line**. Together with the formulae on page 82, these are called the *SUVAT* formulae.

1 $v^2 = u^2 + 2as$ **2** $s = ut + \frac{1}{2}at^2$ **3** $s = vt - \frac{1}{2}at^2$

For a definition of what each letter represents, look at the diagram on page 82.

Worked example

A train moves along a straight track with constant acceleration. Three telegraph poles are set at equal intervals beside the track at points A, B and C, where $AB = 50$ m and $BC = 50$ m. The front of the train passes A with speed 22.5 m s^{-1}, and 2 s later it passes B. Find

(a) the acceleration of the train   **(3 marks)**

$s = 50$, $u = 22.5$, $v = ?$, $a = ?$, $t = 2$

22.5 m s^{-1} $t = 2$

A — 50 m — B — 50 m — C

$s = ut + \frac{1}{2}at^2$

$50 = 22.5 \times 2 + \frac{1}{2} \times a \times 2^2$

$50 = 45 + 2a$

$a = 2.5$ m s^{-2}

(b) the speed of the front of the train when it passes C **(3 marks)**

$s = 100$, $u = 22.5$, $v = ?$, $a = 2.5$, $t = ?$

$v^2 = u^2 + 2as$

$v^2 = 22.5^2 + 2 \times 2.5 \times 100$

$\quad = 1006.25$

$v = 31.72... = 31.7$ m s^{-1} (3 s.f.)

(c) the time that elapses from the instant the front of the train passes B to the instant it passes C.   **(4 marks)**

$s = 50$, $u = ?$, $v = 31.7214...$, $a = 2.5$, $t = ?$

$s = vt - \frac{1}{2}at^2$

$50 = 31.72... \times t - \frac{1}{2} \times 2.5 \times t^2$

$1.25t^2 - 31.72... \, t + 50 = 0$

$t = \dfrac{-(-31.72...) \pm \sqrt{(-31.72...)^2 - 4 \times 1.25 \times 50}}{2.5}$

$t = 1.688...$ or $t = 23.688...$

The time elapsed is 1.69 s. (3 s.f.)

Units

You need to make sure that your measurements are in the correct units.

✓ t (time) is measured in seconds

✓ s (displacement) is measured in metres

✓ u and v (velocity) is measured in m s^{-1}

m s^{-1} means m/s or 'metres per second'

✓ a (acceleration) is measured in m s^{-2}

m s^{-2} means m/s^2, 'metres per second squared', or 'metres per second per second'

Problem solved!

In part (c) you get two solutions to the quadratic equation. You know the train is travelling forwards at B and accelerating. It will take less than 2 seconds to travel from B to C, so choose $t = 1.688...$

You will need to use problem-solving skills throughout your exam – **be prepared!**

Now try this

A boat travels in a straight line with constant deceleration, between two buoys A and B, which are a distance 300 m apart. The boat passes buoy A with initial speed 16 m s^{-1}, and passes buoy B 30 seconds later. Find

(a) the deceleration of the boat   **(3 marks)**

(b) the speed of the boat as it passes B **(4 marks)**

(c) the distance from B at which the boat comes to rest.   **(4 marks)**

Motion under gravity

If you ignore **air resistance**, you can model an object moving freely under gravity as a particle with **constant downward acceleration**. This means you can use the constant acceleration formulae given on pages 82 and 83.

Up and down

Here are some useful facts about particles moving vertically under gravity.

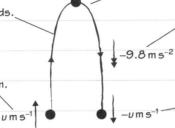

If the particle is thrown upwards and takes t seconds to reach its **maximum height** then it returns to its starting position after $2t$ seconds.

At its **maximum height**, a vertically moving particle has **zero** velocity.

$0\,\text{m s}^{-1}$

You can choose which direction is positive in a question. You usually want UP to be the positive direction.

$-9.8\,\text{m s}^{-2}$

UP is the positive direction, so the acceleration due to gravity is **negative**. It remains **constant** throughout. Show acceleration with a double arrow so you don't confuse it with velocity.

$u\,\text{m s}^{-1}$ $-u\,\text{m s}^{-1}$

When the particle returns to its initial position, it has the **same speed** as it started with but in the **opposite direction**.

Worked example

At time $t = 0$, a particle is projected vertically upwards with speed $u\,\text{m s}^{-1}$ from a point 10 m above the ground. At time T seconds, the particle hits the ground with speed $17.5\,\text{m s}^{-1}$. Find

(a) the value of u **(3 marks)**

$s = -10,\ u = ?,\ v = -17.5,\ a = -9.8,\ \cancel{t = T}$

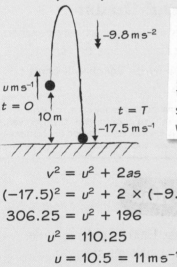

The **positive** direction is **up**, so a and v are **negative**. The finishing position is 10 m **below** the starting position, so s is negative as well.

$v^2 = u^2 + 2as$

$(-17.5)^2 = u^2 + 2 \times (-9.8) \times (-10)$

$306.25 = u^2 + 196$

$u^2 = 110.25$

$u = 10.5 = 11\,\text{m s}^{-1}\ (2\text{ s.f.})$

(b) the value of T.   **(4 marks)**

$\cancel{s = -10},\ u = 10.5,\ v = -17.5,\ a = -9.8,\ t = T$

$v = u + at$

$-17.5 = 10.5 + (-9.8) \times T$

$9.8T = 28$

$T = 2.857... = 2.9\,\text{s}\ (2\text{ s.f.})$

You've used $g = 9.8\,\text{m s}^{-2}$ so round to 2 s.f.

Accuracy and gravity

Unless you're told otherwise, you should use this value for the acceleration due to gravity:

$$g = 9.8\,\text{m s}^{-2}$$

This value for g is correct to **2 significant figures**.

This means that if you use $g = 9.8\,\text{m s}^{-2}$ in your calculation you should give your final answer correct to 2 significant figures.

Don't use rounded values in calculations. In the Worked example on the left, you can give the answer to part (a) as $11\,\text{m s}^{-1}$, but you still need to use $u = 10.5$ in part (b).

Now try this

A diver projects herself upwards from a diving platform with a speed of $3.5\,\text{m s}^{-1}$. The platform is 15 m above the water. Modelling the diver as a particle moving freely under gravity, find

(a) the greatest height above the water reached by the diver   **(4 marks)**

(b) the speed with which the diver hits the water   **(3 marks)**

(c) the total time from when the diver leaves the platform to when she hits the water.   **(3 marks)**

Forces

A force acting on an object has **direction** and **magnitude**. The units of force are **newtons** (N). 1 newton is the force needed to accelerate a 1 kg object at a rate of $1\,ms^{-2}$. Because of this, the units of force can be written as $kg\,m\,s^{-2}$.

$F = ma$

$F = ma$ is sometimes called the **equation of motion**. In words it is:

force (N) = mass (kg) × acceleration ($m\,s^{-2}$)

You need to remember $F = ma$. It is not in the formulae booklet.

This 4 kg block is resting on a smooth surface. If it is acted on by a force of 20 N it will accelerate at a rate of $5\,m\,s^{-2}$.

5 m s⁻²

4 kg → 20 N

Resultant force

If there is more than one force acting on a particle you can find the **resultant** in any given direction.

18 000 N

1 000 N ← → 5 500 N

18 000 N

This boat is accelerating. The vertical forces have the same magnitude so their resultant is **zero**. The resultant force in the horizontal direction is $5500 - 1000 = 4500\,N$.

A car of mass 1000 kg is towing a caravan of mass 750 kg along a straight horizontal road. The caravan is connected to the car by a tow-bar which is parallel to the direction of motion of the car and the caravan. The tow-bar is modelled as a light rod. The engine of the car provides a constant driving force of 3200 N. The resistances to the motion of the car and the caravan are modelled as constant forces of magnitude 800 newtons and R newtons respectively.

Given that the acceleration of the car and the caravan is $0.88\,m\,s^{-2}$,

(a) show that $R = 860$　　　　(**3 marks**)

0.88 m s⁻²

RN ← [750 kg] T T [1000 kg] → 3200 N

800 N

Using $F = ma$ for the whole system:

$$3200 - 800 - R = (750 + 1000) \times 0.88$$
$$2400 - R = 1540$$
$$R = 860\,N$$

(b) find the tension in the tow-bar.　　(**3 marks**)

Using $F = ma$ for the caravan only:

$$T - R = 750 \times 0.88$$
$$T - 860 = 660$$
$$T = 1520\,N$$

For part **(b)** you need to consider either the caravan or the car on its own.

Drawing a large, well-labelled diagram will help you see what is going on. For part **(a)** consider the caravan and the car as a **single system**. The resultant force acting on the system is $3200 - 800 - R$, and the combined mass is $750 + 1000$.

Tension and thrust

✓ **Tension** is a force which will tend to **stretch** a rod, spring or string.

✓ **Thrust** is a force which will tend to **compress** a rod.

When a car accelerates it produces a tension in the tow-bar which in turn accelerates the caravan. If the car brakes, it produces a thrust in the tow-bar which decelerates the caravan.

A car of mass 1200 kg pulls a trailer of mass 400 kg along a straight horizontal road using a light tow-bar. The resistances to motion of the car and trailer have magnitudes 500 N and 200 N respectively. The engine of the car produces a constant driving force of magnitude 1500 N. Find

(a) the acceleration of the car and trailer　　　　(**3 marks**)

(b) the magnitude of the tension in the tow-bar.　　　　(**3 marks**)

Forces as vectors

Force is a **vector quantity**. This means it has **direction** as well as **magnitude**. You can represent forces as **column vectors** or using **i-j notation**. Have a look at page 33 for a reminder about these.

Finding resultants

To find the **resultant** of forces given as vectors, you need to **add** the vectors. You can add the i components and add the j components separately. This particle is being acted on by two forces.

$T_2 = (4i - 3j)\,N$

$T_1 = (-5j)\,N$

$T_R = T_1 + T_2$

The resultant force acting on the particle is $T_R = T_1 + T_2 = (-5j) + (4i - 3j) = (4i - 8j)\,N$

So it has magnitude $|T_R| = \sqrt{4^2 + 8^2} = 8.94\,N$

Worked example

Three forces F_1, F_2 and F_3 acting on a particle P are given by

$F_1 = (7i - 9j)\,N \quad F_2 = (5i + 6j)\,N \quad F_3 = (pi + qj)\,N$

where p and q are constants.

Given that P is in equilibrium,

(a) find the value of p and the value of q.
(3 marks)

$F_1 + F_2 + F_3 = 0$

$(7i - 9j) + (5i + 6j) + (pi + qj) = 0$

i components: $7 + 5 + p = 0$, so $p = -12$

j components: $-9 + 6 + q = 0$, so $q = 3$

The force F_3 is now removed. The resultant of F_1 and F_2 is R. Find

(b) the magnitude of R **(2 marks)**

$R = F_1 + F_2$

$\quad = (7i - 9j) + (5i + 6j)$

$\quad = (12i - 3j)\,N$

$|R| = \sqrt{12^2 + (-3)^2} = 12.4\,N$ (3 s.f.)

(c) the angle, to the nearest degree, that the direction of R makes with j. **(3 marks)**

$R = (12i - 3j)\,N$

$\tan\theta = \dfrac{3}{12}$, so $\theta = 14°$ (nearest degree)

So the angle between j and R is $90° + 14° = 104°$

For part (c), draw a sketch to make sure you calculate the correct angle.

Equilibrium

If a particle is in equilibrium, the resultant force acting on it is ZERO. The vectors of any forces acting on it will form a closed **triangle**.

For forces acting on a particle in equilibrium:

✓ the **i** components add up to zero

✓ the **j** components add up to zero.

Be careful with **accuracy** in vectors questions.

- Give **vectors** exactly in terms of **i** and **j**.
- Give **magnitudes** as surds, or to 3 s.f.
- Give **angles** correct to the **nearest degree**.

Now try this

A particle is acted upon by two forces F_1 and F_2, given by $F_1 = (2i - 6j)\,N$ and $F_2 = (3i + kj)\,N$.

Given that the resultant R of the two forces acts in a direction parallel to the vector $(i - j)$,

(a) find the value of k. **(4 marks)**

A third force F_3, given by $F_3 = (pi + qj)\,N$ is applied to the particle, so that it now lies in equilibrium.

(b) Find the value of p and the value of q.
(3 marks)

Motion in 2D

Acceleration is a vector quantity that can be written as a column vector or using **i-j** notation. You can use this to describe motion in two dimensions.

Equation of motion

You can write equations of motion involving forces and accelerations written as vectors:

$$\mathbf{F} = m\mathbf{a}$$

F is the force vector in newtons

a is the acceleration vector in ms^{-2}

m is the mass in kg.

Mass is a **scalar** quantity. It has **magnitude** but **no direction**. This means that the direction of the force is the same as the direction of the acceleration.

$\mathbf{a} = (3\mathbf{i} - \mathbf{j})\,ms^{-2}$

7 kg

$\mathbf{F} = (21\mathbf{i} - 7\mathbf{j})\,N$

This particle is accelerating from rest at a rate of $\mathbf{a} = (3\mathbf{i} - \mathbf{j})\,ms^{-2}$. This acceleration is produced by a force of $\mathbf{F} = 7(3\mathbf{i} - \mathbf{j}) = (21\mathbf{i} - 7\mathbf{j})\,N$.

Worked example

A hockey ball of mass 160 g is at rest on a smooth flat field. It is pushed by a hockey stick with a constant force of $(-24\mathbf{i} + 45\mathbf{j})\,N$.

(a) Find the acceleration of the hockey ball.
(1 mark)

$$\mathbf{F} = m\mathbf{a}$$
$$(-24\mathbf{i} + 45\mathbf{j}) = 0.16\mathbf{a}$$
$$\mathbf{a} = \frac{1}{0.16}(-24\mathbf{i} + 45\mathbf{j})$$
$$= (-150\mathbf{i} + 281.25\mathbf{j})\,ms^{-2}$$

(b) Given that the ball leaves the hockey stick with a velocity of $44\,ms^{-1}$, find the length of time that the stick is in contact with the ball.
(3 marks)

$$|\mathbf{a}| = \sqrt{150^2 + 281.25^2} = 318.75\,ms^{-2}$$
$$v = u + at$$
$$44 = 0 + 318.75t$$
$$t = 0.138\,s \text{ (3 s.f.)}$$

(c) Criticise this model in respect of:

(i) the field

(ii) the motion of the ball. **(2 marks)**

(i) A hockey field is unlikely to be smooth so there would be some resistance due to friction.

(ii) The force applied by the hockey stick is unlikely to be constant.

You have been given the force as a vector, so you need to give the acceleration in vector form. Use the vector equation of motion $\mathbf{F} = m\mathbf{a}$.

In books vectors are often written in bold type. In your exam you can use a wiggly line underneath the letters to denote a vector quantity.

The ball starts **at rest** and the force acts in a straight line, so the ball will accelerate uniformly in a straight line. You can use the **suvat** equations, taking the **magnitude** of the acceleration as a.

Criticising models

You might be asked to criticise or refine a model in your exam. You should think about the real-life situation and compare this to the model. Try to identify elements of the model that are unrealistic. In real life:

☑ surfaces are rarely smooth

☑ forces and accelerations are rarely constant

☑ objects have dimension and are subject to air resistance and rotation

☑ strings and rods may deform.

Now try this

1 A force of $\begin{pmatrix} 24 \\ -10 \end{pmatrix}$ N acts on a particle of mass 8 kg.

(a) Find the acceleration of the particle, giving your answer as a column vector. **(1 mark)**

(b) Find the magnitude and bearing of the acceleration of the particle. **(4 marks)**

2 A particle of mass 0.6 kg is being acted on by three forces $\mathbf{F_1}\,N$, $\mathbf{F_2}\,N$ and $\mathbf{F_3}\,N$, where

$\mathbf{F_1} = (7\mathbf{i} - 2\mathbf{j})$ $\mathbf{F_2} = (p\mathbf{i} - \mathbf{j})$ $\mathbf{F_3} = q\mathbf{j}$

Given that the particle is accelerating at $(3\mathbf{i} + 6\mathbf{j})\,ms^{-2}$, find the value of p and the value of q. **(3 marks)**

Pulleys

A pulley is used to connect two particles. The particles interact through the **tension** in the string.

Worked example

Two particles A and B have masses $0.5\,\text{kg}$ and $0.2\,\text{kg}$ respectively. The particles are attached to the ends of a light inextensible string which passes over a smooth pulley. Both particles are held, with the string taut, at a height of $2\,\text{m}$ above the floor. The particles are released from rest and in the subsequent motion B does not reach the pulley. Find

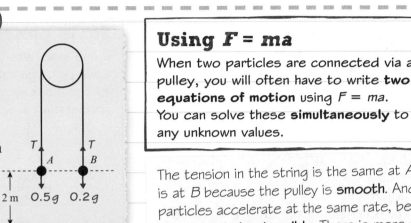

(a) the tension in the string immediately after the particles are released **(6 marks)**

$R(\downarrow)$: Using $F = ma$ for particle A:

$0.5g - T = 0.5a$ ①

$R(\uparrow)$: Using $F = ma$ for particle B:

$T - 0.2g = 0.2a$ ②

$2 \times ①$: $g - 2T = a$

$-5 \times ②$: $\underline{5T - g = a}$

$\qquad\qquad 2g - 7T = 0$

$\qquad\qquad T = \dfrac{2g}{7} = 2.8\,\text{N}$

(b) the acceleration of A immediately after the particles are released **(2 marks)**

Substitute $T = 2.8\,\text{N}$ into ①:

$0.5g - 2.8 = 0.5a$ so $a = 4.2\,\text{ms}^{-2}$

(c) the speed of A when it hits the ground. **(3 marks)**

$s = 2,\ u = 0,\ v = ?,\ a = \dfrac{3g}{7},\ t = ?$

$v^2 = u^2 + 2as$

$\quad = 0^2 + 2 \times \dfrac{3g}{7} \times 2 = \dfrac{12g}{7}$

$v = 4.0987\ldots = 4.1\,\text{ms}^{-1}$ (2 s.f.)

Using $F = ma$

When two particles are connected via a pulley, you will often have to write **two equations of motion** using $F = ma$. You can solve these **simultaneously** to find any unknown values.

The tension in the string is the same at A as it is at B because the pulley is **smooth**. And both particles accelerate at the same rate, because the string is **inextensible**. There is more on **modelling assumptions** like this on page 80.

Problem solved!

Follow these steps for parts (a) and (b).

1. Label your diagram with the tension in the string, and the weight of both particles.
2. Write an equation of motion for each particle. You can resolve up or down for each particle, but remember that the acceleration acts in the opposite direction for B as it does for A.
3. Solve your two equations of motion simultaneously to find T and a.

You will need to use problem-solving skills throughout your exam – **be prepared!**

For part (c), you have to use the formulae for constant acceleration. Have a look at pages 82 and 83 for a reminder.

After the string is cut, particle B behaves as an object moving freely under gravity. Have a look at page 84 for more on this.

Now try this

Two particles A and B have masses $4\,\text{kg}$ and $m\,\text{kg}$ respectively. They are connected by a light inextensible string which passes over a smooth light fixed pulley. The system is released from rest, and A descends with acceleration $0.2g$.

(a) Find the tension in the string as A descends. **(3 marks)**

(b) Find the value of m. **(3 marks)**

After $1\,\text{s}$, the string is cut and particle B moves vertically under gravity.

(c) Find the time after the string is cut at which particle B returns to its initial position. **(9 marks)**

A (4 kg) B (m kg)

Connected particles

When you are solving problems involving connected objects, you need to decide when to consider the **whole system** and when to consider each object **individually**.

One end of a light inextensible string is attached to a block A of mass 20 kg, which is on a rough horizontal table. The other end of the string is attached

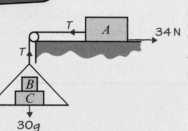

via a pulley to a light scale pan which carries two blocks B and C. The mass of block B is 8 kg and the mass of block C is 22 kg. The system is released from rest.

The resistance to motion of A from the rough table has constant magnitude 34 N. Find:

(a) (i) the acceleration of the scale pan

(ii) the tension in the string.　　**(8 marks)**

(i) Scale pan: $30g - T = 30a$ ①

Block A: $T - 34 = 20a$ ②

① + ②: $30g - 34 = 50a$

$260 = 50a$

$a = 5.2\,\text{ms}^{-2}$

(ii) Substitute into ②:

$T - 34 = 20 \times 5.2$

$T = 138 = 140\,\text{N (2 s.f.)}$

(b) the magnitude of the force exerted on block B by block C.　　**(3 marks)**

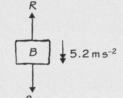

$8g - R = 8 \times 5.2$

$R = 36.8\,\text{N}$

The force exerted on B by C is 37 N (2 s.f.).

If you have to answer questions about **rough surfaces** in your AS exam the magnitude of the frictional force will be constant. It always acts so as to **oppose** the direction of motion.

Scale pans

If one block sits on top of another in a scale pan, you can find the **force exerted** by one block on the other. This scale pan is accelerating at $2.2\,\text{ms}^{-2}$.

On its own, the green block has **two** forces acting on it: its weight, and the **reaction** exerted on it by the orange block. An equation of motion for the green block (resolving in the upwards direction) is

$$R - 3g = 3 \times 2.2$$

So $R = 36\,\text{N}$. Because the two blocks are not moving relative to each other they exert **equal** and **opposite** forces on each other, so the force exerted by the green block on the orange block is the same: 36 N.

Write separate equations of motion for A and for the whole scale pan, then solve simultaneously to find the tension and the acceleration. For part (b) you need to consider either block B or block C on its own.

You will need to use problem-solving skills throughout your exam – **be prepared!**

A block R of mass 0.8 kg is connected by means of a light inextensible string to a light scale pan, which carries two blocks P and Q. P and Q have masses 0.2 kg and m kg respectively. The system is released from rest and block R accelerates upwards at a rate of $1.5\,\text{ms}^{-2}$. Find

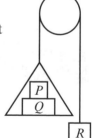

(a) the value of m　　**(6 marks)**

(b) the magnitude of the force exerted on block P by block Q.　　**(3 marks)**

Combining techniques

You might need to answer questions involving forces, and *suvat* equations.

Worked example

A box, *P*, of mass 8 kg is held at rest on a rough platform. It is attached to another box, *Q*, of unknown mass by means of a light inextensible string that runs over a smooth pulley. Box *Q* hangs freely a distance of 2 m above a horizontal floor.

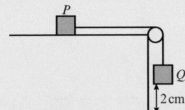

The system is released from rest with the string taut. The frictional force between *P* and the table is modelled as a constant force of magnitude 18.5 N and the boxes are modelled as particles.

(a) Given that box *Q* hits the floor after 2.5 seconds, use the model to find the mass of box *Q*. **(8 marks)**

$s = 2, u = 0, \cancel{v = ?}, a = ?, t = 2.5$
$s = ut + \frac{1}{2}at^2$
$2 = 0 + \frac{1}{2}a \times 2.5^2$
$a = 0.64 \text{ ms}^{-2}$

Let mass of *Q* be m kg and tension in string be *T*.
Equations of motion:
Q: $mg - T = m \times 0.64$ ①
P: $T - 18.5 = 8 \times 0.64$ ②
Solving ① and ② simultaneously:
$mg - 18.5 = 0.64m + 5.12$
$m = \dfrac{5.12 + 18.5}{9.8 - 0.64}$
$= 2.6 \text{ kg (2 s.f.)}$

Following the experiment, box *Q* is weighed and is found to have a mass of 3 kg.

(b) In light of this information:
  (i) comment on the validity of using this model to find the mass of box *Q*
  (ii) suggest one possible improvement to the model. **(2 marks)**

(i) The model is not valid. The actual mass of box *Q* was larger. This could be because in reality the pulley is not smooth, or because air resistance was not considered.
(ii) The resistances to motion could be modelled as a variable force.

Golden rule

If the resultant force acting on a particle is constant, then its acceleration will be constant. This means you can apply the *suvat* formulae to the motion of the particle.

You know the time taken for box *Q* to hit the floor and the distance fallen, so you can use the appropriate *suvat* formula to find the acceleration of box *Q*.

Problem solved!

In your AS exam, any resistances to the motion of an object (such as friction or air resistance) will be modelled as being **constant**. In real life, air resistance increases as the speed of an object increases. If you are asked to suggest an improvement to a model involving a constant resistance, you can suggest that the resistance is modelled as a **variable force** instead.

> You will need to use problem-solving skills throughout your exam – **be prepared!**

Now try this

A trailer *A* of mass m_1 kg is being pulled along a rough horizontal road by a truck *B* of mass m_2 kg. The truck and the trailer are connected by means of a tow rope, which is modelled as a taut, light inextensible string.

The total resistances to motion experienced by the trailer and the truck are modelled as being constant forces of magnitude 1000 N and 1800 N respectively. The truck generates a driving force of 4000 N which causes the truck and trailer to accelerate at 0.4 ms⁻².

At the point when the truck and trailer are travelling at 14 ms⁻¹, the tow rope breaks, causing the trailer to come to rest in a time of 2.8 seconds.

Find the mass of the trailer and the mass of the truck. **(9 marks)**

Variable acceleration 1

You can write displacement, velocity and acceleration as **functions of time**. This allows you to model the motion of an object which is moving in a straight line with variable acceleration.

Worked example

A pipe-inspection robot travels along a straight section of sewer pipe. At time t minutes the distance, s metres, of the robot from its starting position is given by

$$s = \frac{t^3 - 80\,t^2 + 1600t}{50}, 0 \leqslant t \leqslant 40$$

(a) Sketch a distance–time graph for the motion of the robot. **(3 marks)**

$$s = \frac{1}{50}t(t^2 - 80t + 1600)$$

$$= \frac{1}{50}t(t - 40)^2$$

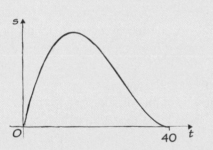

(b) Find the maximum distance of the robot from its starting position. **(6 marks)**

$$\frac{ds}{dt} = \frac{1}{50}(3t^2 - 160t + 1600)$$

$$0 = \frac{1}{50}(3t - 40)(t - 40)$$

$$t = \frac{40}{3} \text{ or } t = 40$$

When $t = \frac{40}{3}$,

$$s = \frac{1}{50}\left(\left(\frac{40}{3}\right)^3 - 80\left(\frac{40}{3}\right)^2 + 1600\left(\frac{40}{3}\right)\right)$$

$$= 189.62...$$

The maximum distance of the robot from its starting point is 190 m (3 s.f.)

You need to know more information about how the function behaves before you can sketch the distance–time graph. You can take out a factor of t and then factorise the quadratic factor. The graph of $s = \frac{1}{50}t(t - 40)^2$ is a cubic graph with a positive coefficient of t^3. It crosses the t-axis at $t = 0$ and touches it at $t = 40$:

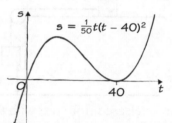

There is more about sketching graphs of cubic functions on page 12.

The model provided is only valid for $0 \leqslant t \leqslant 40$. Make sure you only sketch the section of the graph for these values of t.

From your sketch you can see that $t = 40$ represents a minimum value, so reject that solution.

Maxima and minima

You can use differentiation to find any maximum and minimum values of a function of time. To find the maximum value of $s = f(t)$, differentiate and set $f'(t) = 0$. Solve to find t and then substitute this value back into $f(t)$.

The derivative of the displacement function represents the **velocity**. There is more about this on the next page. There is more about differentiating and finding minima and maxima on pages 39 and 40.

Now try this

A particle travels along the x-axis such that its distance, x m, from the origin at time t seconds is given by:

$$x = t^4 - 20t^3 + 96t^2, 0 \leqslant t \leqslant 8$$

(a) Show that x is non-negative for all values in the given domain of t. **(3 marks)**

(b) Find the maximum distance of the particle from the origin, and justify that it is a maximum. **(8 marks)**

Have a look at page 12 for a reminder about how **quartic** functions behave.

Variable acceleration 2

Velocity is the rate of change of displacement, and acceleration is the rate of change of velocity.

$$v = \frac{ds}{dt} \text{ and } a = \frac{dv}{dt} = \frac{d^2s}{dt^2}$$

Integration is the **inverse** of differentiation, so you can integrate an expression for the velocity to find the displacement, or you can integrate an expression for the acceleration to find the velocity.

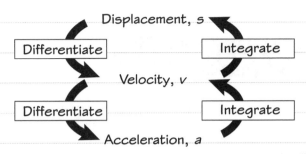

Displacement, s

Differentiate   Integrate

Velocity, v

Differentiate   Integrate

Acceleration, a

Worked example

A particle P moves along the x-axis, passing the origin at time $t = 0$, with velocity 17.6 m s^{-1} in the direction of x increasing.

After a time t seconds, the acceleration of P in the direction of x increasing is $4 - 0.2t \text{ m s}^{-2}$, $t \geq 0$.

Find

(a) the time at which P comes
    instantaneously to rest.   **(6 marks)**

$v = \int (4 - 0.2t)\, dt$

$= 4t - 0.1t^2 + c$

When $t = 0$, $v = 17.6$

$17.6 = 4(0) - 0.1(0)^2 + c \Rightarrow c = 17.6$

So $v = 4t - 0.1t^2 + 17.6$

Set $v = 0$ and solve:

$-0.1t^2 + 4t + 17.6 = 0$

$t = -4$ or $\underline{t = 44}$

P comes to rest after 44 seconds.

(b) the distance travelled by the particle in
    the first two seconds of its motion.
                                 (4 marks)

$s = \int_0^2 (4t - 0.1t^2 + 17.6)\, dt$

$= \left[2t^2 - \frac{1}{30}t^3 + 17.6t \right]_0^2$

$= \left(2(2)^2 - \frac{1}{30}(2)^3 + 17.6(2) \right) - 0$

$= 42.9 \text{ m (3 s.f.)}$

Completing the function

An indefinite integral always produces a function with a **constant of integration**. You need to find this constant in order to use the function to solve numerical problems. When you integrate to find velocity or displacement, you might need to use a given value to find the constant of integration. If this is the value when $t = 0$ it is called an **initial condition**.

There is more about finding functions by integrating on page 43.

Problem solved!

Follow these steps for part (a):

1. Integrate the expression for the acceleration of P to find an expression for the velocity.

2. Use the initial conditions to find the value of the constant of integration.

3. Set the velocity equal to 0 and solve. The model is only valid for $t \geq 0$ so you can reject the negative root of the equation.

You will need to use problem-solving skills throughout your exam – **be prepared!**

You can use **definite integration** to find the distance travelled between two given times. This is the same as the area under the velocity–time graph between $t = 0$ and $t = 2$.

Now try this

A target in a fairground shooting range moves along a straight line. At time t seconds, its distance from a fixed point O is x m and its velocity, $v \text{ m s}^{-1}$, is given by: $v = 0.1\sqrt{t} - 0.2t$, $0 \leq t \leq 6$

When $t = 0$, $x = 4$.

(a) Find the acceleration of the target when $t = 4$.
                                          (4 marks)

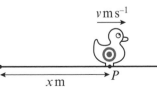

$v \text{ m s}^{-1}$

O ←———— x m ————→ P

(b) Find an expression for x, giving your
    answer in the form $x = f(t)$ **(4 marks)**

(c) Find the value of x when $t = 5$.   **(1 mark)**

Deriving *suvat* equations

You can use **integration** to derive the *suvat* equations for motion with constant acceleration.

Worked example

A particle moves in a straight line with constant acceleration, $A\,\mathrm{m\,s^{-2}}$. At time t seconds the velocity of the particle is $v\,\mathrm{m\,s^{-1}}$ and its displacement relative to a fixed origin is $s\,\mathrm{m}$.

Given that the initial velocity of the particle is $U\,\mathrm{m\,s^{-1}}$ and its initial displacement is $0\,\mathrm{m}$, prove that

(a) $v = U + At$ **(3 marks)**

$v = \int A\,dt$

$\quad = At + c$

When $t = 0$, $v = U \Rightarrow c = U$

So $v = At + U = U + At$ as required

(b) $s = Ut + \frac{1}{2}At^2$ **(3 marks)**

$s = \int v\,dt = \int(U + At)dt$

$\quad = Ut + \frac{1}{2}At^2 + c$

When $t = 0$, $s = U \Rightarrow c = 0$

So $s = Ut + \frac{1}{2}At^2$ as required

Integrate, then use the given initial conditions to find the constants of integration. Remember to write your final answers in the form asked for in the question.

> In this example capital letters are used to represent letters that can be treated as constants when you integrate.

Worked example

Given that

$\quad v = u + at$ ①

$\quad s = ut + \frac{1}{2}at^2$ ②

prove that $v^2 = u^2 + 2as$. You may not make use of any formulae other than those given above.   **(3 marks)**

Squaring both sides of ①:

$v^2 = (u + at)^2$

$\quad = u^2 + 2uat + a^2t^2$

$\quad = u^2 + 2a(ut + \frac{1}{2}at^2)$

Substituting ② into the above:

$v^2 = u^2 + 2as$ as required.

You can use the two *suvat* formulae in the example above left to derive the other *suvat* formulae. Have a look at pages 82 and 83 for a reminder about the five *suvat* formulae.

Using a velocity–time graph

You can also derive the *suvat* formulae from a velocity–time graph. The diagram shows a velocity–time graph for a particle accelerating constantly from velocity $U\,\mathrm{m\,s^{-1}}$ to velocity $V\,\mathrm{m\,s^{-1}}$ in time T seconds. The acceleration, $a\,\mathrm{m\,s^{-2}}$, of the particle is equal to the gradient of the graph:

$a = \dfrac{V - U}{T}$, so $V = U + aT$

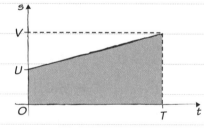

Now try this

1   A particle moves in a straight line along the x-axis. At time t seconds, its distance from the origin is given by

$\quad s = 10t - kt^2, t \geqslant 0$

where k is a constant.

(a) Show that the particle is moving with constant acceleration. **(4 marks)**

(b) Given that the particle is instantaneously at rest when $t = 4$, find the value of k. **(3 marks)**

2   Use the velocity–time graph above to show that the distance travelled, $s\,\mathrm{m}$, by the particle in time T seconds is given by

$\quad s = \left(\dfrac{U + V}{2}\right)T$ **(4 marks)**

You are the examiner!

In your exam you might be asked to identify errors in working. You should also be confident **checking your own work**. Each of these students has made key mistakes in their working. Can you spot them all?

1 A stone is thrown vertically upwards with speed $16\,\mathrm{m\,s^{-1}}$ from a point h metres above the ground. The stone hits the ground 4 s later.

Find (a) the value of h **(3 marks)**

(b) the speed of the stone as it hits the ground.   **(3 marks)**

$s = h,\ u = 16,\ v = ?,\ a = 9.8,\ t = 4$

(a) $s = ut + \frac{1}{2}at^2$

$= 16 \times 4 + \frac{1}{2} \times 9.8 \times 4^2$

$= 142.4$

$h = 140$ (2 s.f.)

(b) $v = u + at$

$= 16 + 9.8 \times 4$

$= 55$ (2 s.f.)

2 In this question **i** and **j** represent the unit vectors due east and north respectively. A particle is acted on by two forces, $(2\mathbf{i} - 5\mathbf{j})$ N and $(7\mathbf{i} + \mathbf{j})$ N. Find the magnitude and bearing of the resultant force acting on the particle.   **(4 marks)**

Resultant force $= (2\mathbf{i} - 5\mathbf{j}) + (7\mathbf{i} + \mathbf{j})$

$= (9\mathbf{i} - 4\mathbf{j})$ N

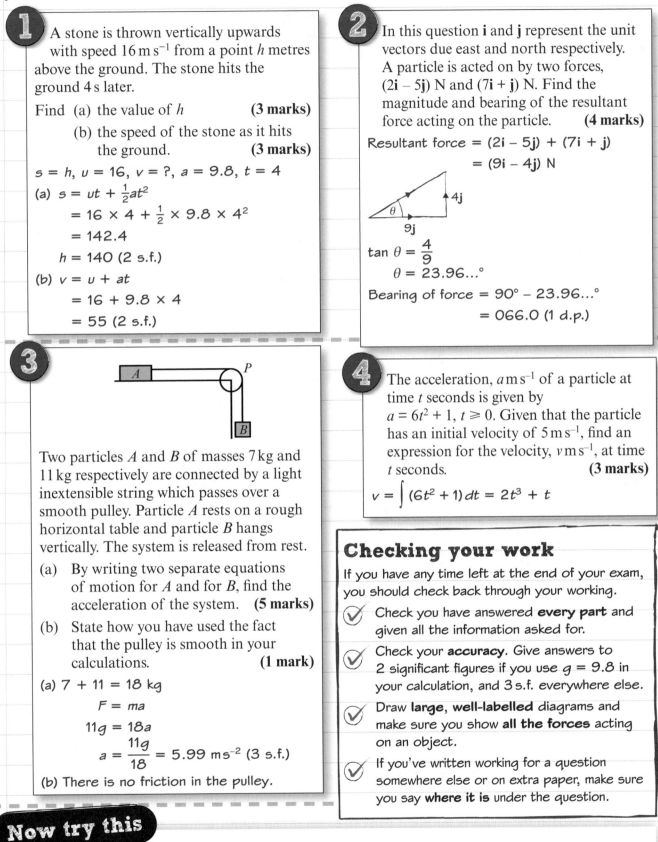

$\tan \theta = \dfrac{4}{9}$

$\theta = 23.96...°$

Bearing of force $= 90° - 23.96...°$

$= 066.0$ (1 d.p.)

3

Two particles A and B of masses 7 kg and 11 kg respectively are connected by a light inextensible string which passes over a smooth pulley. Particle A rests on a rough horizontal table and particle B hangs vertically. The system is released from rest.

(a) By writing two separate equations of motion for A and for B, find the acceleration of the system.   **(5 marks)**

(b) State how you have used the fact that the pulley is smooth in your calculations.   **(1 mark)**

(a) $7 + 11 = 18$ kg

$F = ma$

$11g = 18a$

$a = \dfrac{11g}{18} = 5.99\ \mathrm{ms^{-2}}$ (3 s.f.)

(b) There is no friction in the pulley.

4 The acceleration, $a\,\mathrm{m\,s^{-1}}$ of a particle at time t seconds is given by $a = 6t^2 + 1,\ t \geqslant 0$. Given that the particle has an initial velocity of $5\,\mathrm{m\,s^{-1}}$, find an expression for the velocity, $v\,\mathrm{m\,s^{-1}}$, at time t seconds.   **(3 marks)**

$v = \displaystyle\int (6t^2 + 1)\,dt = 2t^3 + t$

Checking your work

If you have any time left at the end of your exam, you should check back through your working.

✓ Check you have answered **every part** and given all the information asked for.

✓ Check your **accuracy**. Give answers to 2 significant figures if you use $g = 9.8$ in your calculation, and 3 s.f. everywhere else.

✓ Draw **large, well-labelled** diagrams and make sure you show **all the forces** acting on an object.

✓ If you've written working for a question somewhere else or on extra paper, make sure you say **where it is** under the question.

Now try this

Find the mistakes in each student's answer, and write out the correct working for each question. Turn over for the answers.

You are still the examiner!

Before looking at this page, turn back to page 94 and try to spot the key mistakes in each student's working. Use this page to check your answers. The corrections are shown in red and these answers are now 100% correct.

1 A stone is thrown vertically upwards with speed $16\,\mathrm{m\,s^{-1}}$ from a point h metres above the ground. The stone hits the ground $4\,\mathrm{s}$ later.

Find (a) the value of h **(3 marks)**

(b) the speed of the stone as it hits the ground. **(3 marks)**

Taking upwards as the positive direction

$s = -h, u = 16, v = ?, a = -9.8, t = 4$

(a) $s = ut + \frac{1}{2}at^2 = 16 \times 4 - \frac{1}{2} \times 9.8 \times 4^2$

~~$= 16 \times 4 + \frac{1}{2} \times 9.8 \times 4^2$~~ $= -14.4$

~~$= 142.4$~~ $h = 14$ (2 s.f.)

~~$h = 140$ (2 s.f.)~~

(b) $v = u + at$ $= 16 - 9.8 \times 4$

~~$= 16 + 9.8 \times 4$~~ $= -23.2$

~~$= 55.2\,\mathrm{ms^{-1}}$~~   Speed $= 23\,\mathrm{m\,s^{-1}}$ (2 s.f.)

Top tip

Direction is important in **suvat** questions. If your positive direction is up, then acceleration is **negative**.

Revise motion under gravity on page 84.

2 In this question **i** and **j** represent the unit vectors due east and north respectively. Particle A is acted on by two forces, $(2\mathbf{i} - 5\mathbf{j})$ N and $(7\mathbf{i} + \mathbf{j})$ N. Find the magnitude and bearing of the resultant force acting on the particle. **(4 marks)**

Resultant force $= (2\mathbf{i} - 5\mathbf{j}) + (7\mathbf{i} + \mathbf{j})$

$= (9\mathbf{i} - 4\mathbf{j})$ N

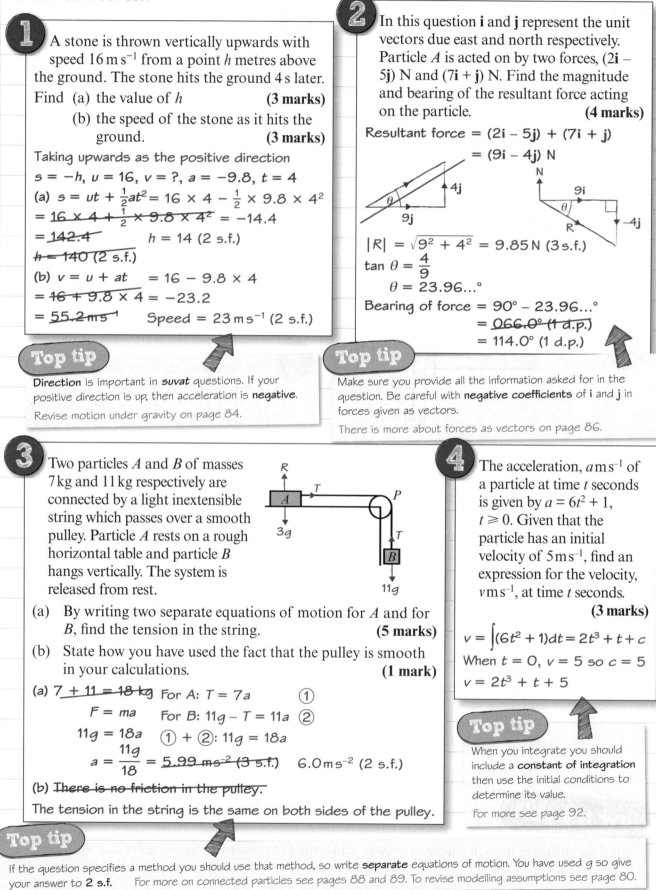

$|R| = \sqrt{9^2 + 4^2} = 9.85\,\mathrm{N}$ (3 s.f.)

$\tan\theta = \dfrac{4}{9}$

$\theta = 23.96...°$

Bearing of force $= 90° - 23.96...°$

~~$= 066.0°$ (1 d.p.)~~

$= 114.0°$ (1 d.p.)

Top tip

Make sure you provide all the information asked for in the question. Be careful with **negative coefficients** of **i** and **j** in forces given as vectors.

There is more about forces as vectors on page 86.

3 Two particles A and B of masses $7\,\mathrm{kg}$ and $11\,\mathrm{kg}$ respectively are connected by a light inextensible string which passes over a smooth pulley. Particle A rests on a rough horizontal table and particle B hangs vertically. The system is released from rest.

(a) By writing two separate equations of motion for A and for B, find the tension in the string. **(5 marks)**

(b) State how you have used the fact that the pulley is smooth in your calculations. **(1 mark)**

(a) ~~$7 + 11 = 18\,\mathrm{kg}$~~ For A: $T = 7a$ ①

$F = ma$ For B: $11g - T = 11a$ ②

$11g = 18a$ ① + ②: $11g = 18a$

$a = \dfrac{11g}{18} = $ ~~$5.99\,\mathrm{ms^{-2}}$ (3 s.f.)~~ $6.0\,\mathrm{m\,s^{-2}}$ (2 s.f.)

(b) ~~There is no friction in the pulley.~~

The tension in the string is the same on both sides of the pulley.

4 The acceleration, $a\,\mathrm{m\,s^{-1}}$ of a particle at time t seconds is given by $a = 6t^2 + 1$, $t \geq 0$. Given that the particle has an initial velocity of $5\,\mathrm{m\,s^{-1}}$, find an expression for the velocity, $v\,\mathrm{m\,s^{-1}}$, at time t seconds. **(3 marks)**

$v = \int(6t^2 + 1)dt = 2t^3 + t + c$

When $t = 0$, $v = 5$ so $c = 5$

$v = 2t^3 + t + 5$

Top tip

When you integrate you should include a **constant of integration** then use the initial conditions to determine its value.

For more see page 92.

Top tip

If the question specifies a method you should use that method, so write **separate** equations of motion. You have used g so give your answer to 2 s.f.   For more on connected particles see pages 88 and 89. To revise modelling assumptions see page 80.

Worked solutions

PURE

1. Index laws

1 $x(4x^{-\frac{1}{2}})^3 = x^1(4^3x^{-\frac{3}{2}})$
$\qquad = 64x^{-\frac{3}{2}+1} = 64x^{-\frac{1}{2}}$

2 $(9y^{10})^{\frac{3}{2}} = 9^{\frac{3}{2}}y^{10\times\frac{3}{2}}$
$\qquad = (\sqrt{9})^3y^{15} = 27y^{15}$

3 $\dfrac{5 + 2\sqrt{x}}{x^2} = \dfrac{5}{x^2} + \dfrac{2x^{\frac{1}{2}}}{x^2}$
$\qquad = 5x^{-2} + 2x^{\frac{1}{2}-2}$
$\qquad = 5x^{-2} + 2x^{-\frac{3}{2}}$

2. Expanding and factorising

1 $(x + 2)(x + 1)^2 = (x + 2)(x^2 + 2x + 1)$
$\qquad = x^3 + 2x^2 + x + 2x^2 + 4x + 2$
$\qquad = x^3 + 4x^2 + 5x + 2$
$b = 4, c = 5, d = 2$

2 $3x^3 - 2x^2 - x = x(3x^2 - 2x - 1)$
$\qquad = x(3x + 1)(x - 1)$

3 $25x^2 - 16 = (5x + 4)(5x - 4)$

3. Surds

1 $(x + \sqrt{3})(x - \sqrt{3}) = x^2 + x\sqrt{3} - x\sqrt{3} - \sqrt{3}\sqrt{3} = x^2 - 3$

2 $\sqrt{98} = \sqrt{49} \times \sqrt{2} = 7\sqrt{2}$

3 (a) $\dfrac{8}{3 + \sqrt{5}} = \dfrac{8(3 - \sqrt{5})}{(3 + \sqrt{5})(3 - \sqrt{5})} = \dfrac{24 - 8\sqrt{5}}{4} = 6 - 2\sqrt{5}$

(b) $\dfrac{4 + \sqrt{5}}{2 - \sqrt{5}} = \dfrac{(4 + \sqrt{5})(2 + \sqrt{5})}{(2 - \sqrt{5})(2 + \sqrt{5})} = \dfrac{13 + 6\sqrt{5}}{-1} = -13 - 6\sqrt{5}$

4. Quadratic equations

1 $\qquad 2(x - 3)^2 + 3x = 14$
$2(x^2 - 6x + 9) + 3x = 14$
$2x^2 - 12x + 18 + 3x = 14$
$\qquad 2x^2 - 9x + 4 = 0$
$\qquad (2x - 1)(x - 4) = 0$
$x = \frac{1}{2}$ or $x = 4$

2 (a) $x^2 - 10x + 7 = (x - 5)^2 - 5^2 + 7$
$\qquad = (x - 5)^2 - 18$
$\qquad p = 1, q = -18$

(b) $x^2 - 10x + 7 = 0$
$\qquad (x - 5)^2 - 18 = 0$
$\qquad (x - 5)^2 = 18$
$\qquad x - 5 = \pm\sqrt{18} = \pm3\sqrt{2}$
$\qquad x = 5 \pm 3\sqrt{2}$
$\qquad a = 5, b = 3$

5. Functions and roots

1 (a) $\qquad x^4 - 3x^2 - 4 = 0$
$\qquad (x^2)^2 - 3x^2 - 4 = 0$
$\qquad (x^2 + 1)(x^2 - 4) = 0$
$x^2 = -1$ or $x^2 = 4$ so $x = 2$ or $x = -2$

(b) $\qquad 8x^6 + 7x^3 - 1 = 0$
$\qquad 8(x^3)^2 + 7x^3 - 1 = 0$
$\qquad (8x^3 - 1)(x^3 + 1) = 0$
$x^3 = -1$ or $x^3 = \frac{1}{8}$ so $x = -1$ or $x = \frac{1}{2}$

(c) $\qquad x + 10 = 7\sqrt{x}$
$\qquad (\sqrt{x})^2 - 7\sqrt{x} + 10 = 0$
$\qquad (\sqrt{x} - 5)(\sqrt{x} - 2) = 0$
$\sqrt{x} = 5$ or $\sqrt{x} = 2$ so $x = 25$ or $x = 4$

2 $\qquad 4\sqrt{x} + x = 3$
$\qquad (\sqrt{x})^2 + 4\sqrt{x} - 3 = 0$
$\qquad (\sqrt{x} + 2)^2 - 4 - 3 = 0$
$\qquad \sqrt{x} + 2 = \pm\sqrt{7}$
$\qquad \sqrt{x} = -2 \pm \sqrt{7}$
so $\sqrt{x} = -2 + \sqrt{7}$ or $x = -2 - \sqrt{7}$
$\qquad x = (-2 + \sqrt{7})^2$
$\qquad = 11 - 4\sqrt{7}$

3 $\qquad \text{f}(x) = 0$
$\qquad x^4 - 4x^2 - 5 = 0$
$\qquad (x^2 - 5)(x^2 + 1) = 0$
$\qquad x^2 = 5$ or $x^2 = -1$
$x \in \mathbb{R}, x < 0$ so the only root is $x = -\sqrt{5}$

6. Sketching quadratics

1 (a)

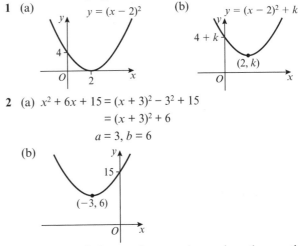

$y = (x - 2)^2$

(b) $y = (x - 2)^2 + k$

2 (a) $x^2 + 6x + 15 = (x + 3)^2 - 3^2 + 15$
$\qquad = (x + 3)^2 + 6$
$\qquad a = 3, b = 6$

(b)

(c) The graph does not intersect the x-axis so the equation has no real solutions.

7. The discriminant

1 $(-2)^2 - 4 \times 3 \times (-5) = 64$

2 $\qquad b^2 - 4ac = 0$
$\qquad 2^2 - 4p \times (-3) = 0$
$\qquad 4 + 12p = 0$
$\qquad p = -\frac{1}{3}$

3 (a) $(k + 5)^2 - 4 \times 1 \times 2k = k^2 + 10k + 25 - 8k$
$\qquad = k^2 + 2k + 25$
$\qquad p = 1, q = 24$

(b) $k^2 + 2k + 25 = (k + 1)^2 - 1^2 + 25$
$\qquad = (k + 1)^2 + 24$

(c) $(k + 1)^2 \geqslant 0$ for all k, so discriminant > 0 for all k, so $\text{f}(x) = 0$ has distinct real roots.

8. Modelling with quadratics

(a) $42\,\text{m}$

(b) $0 = 42 + 0.7d - 0.14d^2$

$\qquad d = \dfrac{-0.7 \pm \sqrt{0.7^2 - 4(-0.14)(42)}}{2(-0.14)}$

$\qquad d = -15$ or $d = 20$

$d > 0$ so the horizontal distance is $20\,\text{m}$.

(c) $h = 42 + 0.7d - 0.14d^2$
$= 42 - 0.14(d^2 - 5d)$
$= 42 - 0.14((d - 2.5)^2 - 2.5^2)$
$= 42.875 - 0.14(d - 2.5)^2$
$0.14(d - 2.5)^2 > 0$ so the maximum height is $42.9\,\text{m}$ (3 s.f.)

9. Simultaneous equations

1   $x + y = 5$ ①
   $x^2 + 2y^2 = 22$ ②
   From ①: $x = 5 - y$
   Substitute into ②: $(5 - y)^2 + 2y^2 = 22$
                 $25 - 10y + y^2 + 2y^2 = 22$
                     $3y^2 - 10y + 3 = 0$
                     $(3y - 1)(y - 3) = 0$
   $y = \frac{1}{3} \Rightarrow x = 5 - \frac{1}{3} = \frac{14}{3}$
   $y = 3 \Rightarrow x = 5 - 3 = 2$
   Solutions: $x = \frac{14}{3}, y = \frac{1}{3}$ and
                $x = 2, y = 3$

2   (a) $y = x + 6$ ①
      $xy - 2x^2 = 7$ ②
      Substituting ① into ②: $x(x + 6) - 2x^2 = 7$
                      $x^2 + 6x - 2x^2 = 7$
                      $-x^2 + 6x - 7 = 0$
                      $x^2 - 6x + 7 = 0$
   (b) $(x - 3)^2 - 3^2 + 7 = 0$
          $(x - 3)^2 = 2$
              $x = 3 \pm \sqrt{2}$
     $x = 3 + \sqrt{2} \Rightarrow y = 9 + \sqrt{2}$
     $x = 3 - \sqrt{2} \Rightarrow y = 9 - \sqrt{2}$
     Solutions: $x = 3 + \sqrt{2}, y = 9 + \sqrt{2}$ and
               $x = 3 - \sqrt{2}, y = 9 - \sqrt{2}$

10. Inequalities

1     $x(x - 5) < 14$
    $x^2 - 5x - 14 < 0$
    $(x - 7)(x + 2) < 0$

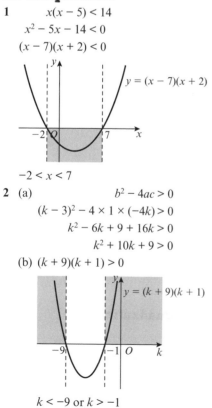

   $-2 < x < 7$

2   (a) $b^2 - 4ac > 0$
      $(k - 3)^2 - 4 \times 1 \times (-4k) > 0$
        $k^2 - 6k + 9 + 16k > 0$
          $k^2 + 10k + 9 > 0$
   (b) $(k + 9)(k + 1) > 0$

   $k < -9$ or $k > -1$

11. Inequalities on graphs

(a), (d)

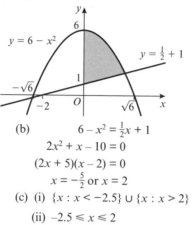

   (b) $6 - x^2 = \frac{1}{2}x + 1$
        $2x^2 + x - 10 = 0$
      $(2x + 5)(x - 2) = 0$
        $x = -\frac{5}{2}$ or $x = 2$
   (c) (i) $\{x : x < -2.5\} \cup \{x : x > 2\}$
       (ii) $-2.5 \leqslant x \leqslant 2$

12. Cubic and quartic graphs

1   (a) $x^3 - 9x = x(x^2 - 9) = x(x + 3)(x - 3)$
   (b)

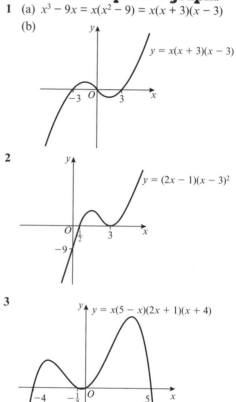

2

3

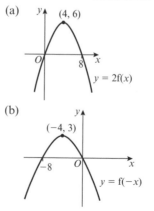

13. Transformations 1

(a)

(b)

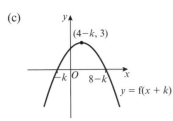

(c)

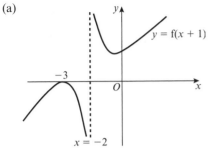

$(4-k, 3)$

$-k \quad O \quad 8-k \quad x$

$y = f(x + k)$

14. Transformations 2

(a)

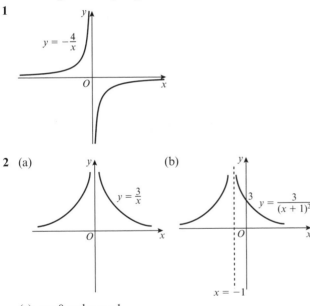

$y = f(x + 1)$

-3

O

$x = -2$

(b) When $x = 0$, $f(x + 1) = f(1)$

$$f(1) = \frac{(1 + 2)^2}{1 + 1} = \frac{9}{2} = 4\frac{1}{2}$$

$(-3, 0)$ and $\left(0, 4\frac{1}{2}\right)$

15. Reciprocal graphs

1

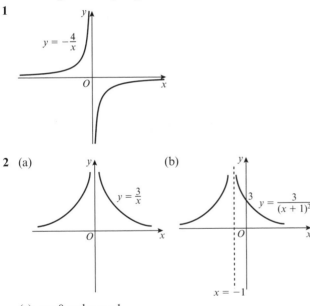

$y = -\frac{4}{x}$

$O \quad x$

2 (a)

$y = \frac{3}{x}$

$O \quad x$

(b)

$3 \quad y = \frac{3}{(x + 1)^2}$

$O \quad x$

$x = -1$

(c) $y = 0$ and $x = -1$

16. Points of intersection

(a)

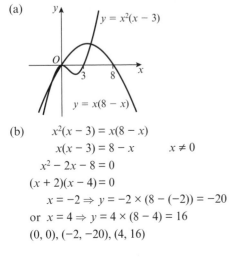

$y = x^2(x - 3)$

O

$3 \quad 8 \quad x$

$y = x(8 - x)$

(b) $\quad x^2(x - 3) = x(8 - x)$

$\quad\quad x(x - 3) = 8 - x \quad\quad x \neq 0$

$\quad x^2 - 2x - 8 = 0$

$\quad (x + 2)(x - 4) = 0$

$\quad\quad x = -2 \Rightarrow y = -2 \times (8 - (-2)) = -20$

or $x = 4 \Rightarrow y = 4 \times (8 - 4) = 16$

$(0, 0)$, $(-2, -20)$, $(4, 16)$

17. Equations of lines

1 $\quad\quad y - y_1 = m(x - x_1)$

$\quad\quad y - (-5) = -\frac{1}{3}(x - 6)$

$\quad\quad\quad 3y + 15 = -x + 6$

$\quad\quad x + 3y + 9 = 0$

2 $\quad m = \frac{y_2 - y_1}{x_2 - x_1} = \frac{11 - 2}{8 - (-4)} = \frac{3}{4}$

$\quad\quad y - y_1 = m(x - x_1)$

$\quad\quad y - 11 = \frac{3}{4}(x - 8)$

$\quad\quad\quad y = \frac{3}{4}x + 5$

3 (a) $\quad 3(1) + 4(5) - k = 0$

$\quad\quad\quad\quad\quad\quad k = 23$

(b) $\quad 3y + 4x - 23 = 0$

$\quad\quad\quad 3y = -4x + 23$

$\quad\quad\quad\quad y = -\frac{4}{3}x + \frac{23}{3}$

$\quad$ Gradient $= -\frac{4}{3}$

18. Parallel and perpendicular

1 (a) $\quad 10 - 3x = 10 - 3(4) = 10 - 12 = -2$

(b) $\quad m = -\frac{1}{-3} = \frac{1}{3}$

$\quad\quad y - y_1 = m(x - x_1)$

$\quad\quad y - (-2) = \frac{1}{3}(x - 4)$

$\quad\quad\quad 3y + 6 = x - 4$

$\quad\quad x - 3y - 10 = 0$

2 $\quad 4x - 5(0) - 1 = 0 \Rightarrow x = \frac{1}{4}$

$\quad A$ is the point $(\frac{1}{4}, 0)$

$\quad 4x - 5y - 1 = 0 \Rightarrow y = \frac{4}{5}x - \frac{1}{5}$

$\quad$ Gradient of $L_1 = \frac{4}{5}$

$\quad$ Gradient of $L_2 = -\frac{5}{4}$

$\quad\quad y - y_1 = m(x - x_1)$

$\quad\quad y - 0 = -\frac{5}{4}(x - \frac{1}{4})$

$\quad\quad\quad y = -\frac{5}{4}x + \frac{5}{16}$

19. Lengths and areas

(a) $\quad x - 2(0) + 6 = 0 \Rightarrow x = -6$

$\quad P$ is $(-6, 0)$

$\quad (0) - 2y + 6 = 0 \Rightarrow y = 3$

$\quad Q$ is $(0, 3)$

$\quad PQ = \sqrt{6^2 + 3^2} = \sqrt{45} = \sqrt{9 \times 5} = 3\sqrt{5}$

(b) $\quad x - 2y + 6 = 0 \Rightarrow y = \frac{1}{2}x + 3$

$\quad$ Gradient of $L_1 = \frac{1}{2}$

$\quad$ Gradient of $L_2 = -2$

$\quad\quad y - 3 = -2(x - 0)$

$\quad\quad\quad y = -2x + 3$

(c) $\quad 0 = -2x + 3 \Rightarrow x = \frac{3}{2}$

$\quad R$ is $(\frac{3}{2}, 0)$

$\quad PR = 6 + \frac{3}{2} = \frac{15}{2}$

$\quad$ Area $= \frac{1}{2} \times \frac{15}{2} \times 3 = \frac{45}{4}$

20. Equation of a circle

(a) $\quad AB = \sqrt{(2 - (-6))^2 + (4 - 0)^2} = \sqrt{80} = 4\sqrt{5}$

(b) $\quad \left(\frac{-6 + 2}{2}, \frac{0 + 4}{2}\right) = (-2, 2)$

(c) $\quad$ Centre $(-2, 2)$, radius $2\sqrt{5}$

$\quad\quad (x - (-2))^2 + (y - 2)^2 = (2\sqrt{5})^2$

$\quad\quad\quad (x + 2)^2 + (y - 2)^2 = 20$

21. Circle properties

(a) $(x - 5)^2 + (y + 2)^2 = 100$

(b) $(-1 - 5)^2 + (6 + 2)^2 = (-6)^2 + 8^2 = 36 + 64 = 100$

So $(-1, 6)$ lies on C.

(c) Radius between $(-1, 6)$ and $(5, -2)$ has gradient

$\dfrac{(-2) - 6}{5 - (-1)} = \dfrac{-8}{6} = -\dfrac{4}{3}$

Tangent is perpendicular to radius, so has gradient $\frac{3}{4}$.

$$y - y_1 = m(x - x_1)$$
$$y - 6 = \tfrac{3}{4}(x + 1)$$
$$4y - 24 = 3x + 3$$
$$3x - 4y + 27 = 0$$

22. Circles and lines

1　$x^2 + (x + 7 + 2)^2 = 45$

$x^2 + x^2 + 18x + 81 = 45$

$x^2 + 9x + 18 = 0$

$(x + 6)(x + 3) = 0$

So $x = -6$ or $x = -3$

Points of intersection are $(-6, 1)$ and $(-3, 4)$.

2　$2x - y + 2 = 0$ so $y = 2x + 2$

Equation of circle:

$(x - k)^2 + y^2 = 4$

So points of intersection satisfy:

$(x - k)^2 + (2x + 2)^2 = 4$

$x^2 - 2kx + k^2 + 4x^2 + 8x + 4 = 4$

$5x^2 + (8 - 2k)x + k^2 = 0$

Two distinct solutions so $b^2 - 4ac > 0$:

$(8 - 2k)^2 - 20k^2 > 0$

$64 - 32k + 4k^2 - 20k^2 > 0$

$64 - 32k - 16k^2 > 0$

$k^2 + 2k - 4 < 0$

Solutions to $k^2 - 2k - 4 = 0$ are $k = -1 \pm \sqrt{5}$

So values of k that satisfy inequality are:

$-1 - \sqrt{5} < \text{k} < -1 + \sqrt{5}$

3　Perpendicular bisector of $(0, 0)$ and $(0, 10)$ is $y = 5$

Perpendicular bisector of $(0, 0)$ and $(8, -6)$ passes through

$(4, -3)$ and has gradient $\dfrac{1}{\left(-\frac{3}{4}\right)} = \dfrac{4}{3}$.

So has equation: $y + 3 = \frac{4}{3}(x - 4)$ or $3y - 4x + 25 = 0$

Substitute $y = 5$ into this equation:

$3(5) - 4x + 25 = 0$ so $x = 10$.

Centre of circle is $(10, 5)$.

Radius of circle is $\sqrt{10^2 + 5^2} = \sqrt{125}$

So equation of circle is $(x - 10)^2 + (y - 5)^2 = 125$

23. The factor theorem

1　(a) $f(2) = 2^3 - 7(2)^2 - 14(2) + 48$

$= 8 - 28 - 28 + 48$

$= 0$

So $(x - 2)$ is a factor.

(b)
2	1	−7	−14	48
	↓	2	−10	−48
	1	−5	−24	0

$f(x) = (x - 2)(x^2 - 5x - 24)$

$= (x - 2)(x - 8)(x + 3)$

2　(a)　　　　　　　　$f(-4) = 0$

$2(-4)^3 - 3(-4)^2 - 65(-4) - a = 0$

$-128 - 48 + 260 - a = 0$

$a = 84$

(b)
−4	2	−3	−65	−84
	↓	−8	44	84
	2	−11	−21	0

$f(x) = (x + 4)(2x^2 - 11x - 21)$

$= (x + 4)(2x + 3)(x - 7)$

24. The binomial expansion

1　(a) $(1 + 3x)^9 = 1^9 + \binom{9}{1} \times 1^8 \times 3x + \binom{9}{2} \times 1^7 \times (3x)^2$

$+ \binom{9}{3} \times 1^6 \times (3x)^3 + \ldots$

$= 1 + 27x + 324x^2 + 2268x^3 + \ldots$

(b) $(2 + 5x)^4 = 2^4 + \binom{4}{1} \times 2^3 \times 5x + \binom{4}{2} \times 2^2 \times (5x)^2$

$+ \binom{4}{3} \times 2 \times (5x)^3 + \ldots$

$= 16 + 160x + 600x^2 + 1000x^3 + \ldots$

(c) $(3 - x)^{12} = 3^{12} + \binom{12}{1} \times 3^{11} \times (-x) + \binom{12}{2} \times 3^{10} \times (-x)^2$

$+ \binom{12}{3} \times 3^9 \times (-x)^3 + \ldots$

$= 531\,441 - 2\,125\,764x + 3\,897\,234x^2$

$- 4\,330\,260x^3 + \ldots$

2　(a) $(2 + kx)^5 = 2^5 + \binom{5}{1} \times 2^4 \times kx + \binom{5}{2} \times 2^3 \times (kx)^2 + \ldots$

$= 32 + 80kx + 80k^2x^2 + \ldots$

(b) $80k = 48$

$k = \frac{3}{5}$

(c) $80k^2 = 80 \times \left(\frac{3}{5}\right)^2 = \frac{144}{5}$

25. Solving binomial problems

1　(a) x^3 term $= \binom{30}{3} \times 1^{27} \times (2x)^3 = 32\,480x^3$

$p = 32\,480$

(b) x^4 term $= \binom{30}{4} \times 1^{26} \times (2x)^4 = 438\,480x^4$

$q = 438\,480$ and $\dfrac{q}{p} = \dfrac{438\,480}{32\,480} = \dfrac{27}{2}$

2　(a) $\left(1 + \frac{x}{4}\right)^{12} = 1^{12} + \binom{12}{1} \times 1^{11} \times \left(\frac{x}{4}\right) + \binom{12}{2} \times 1^{10} \times \left(\frac{x}{4}\right)^2$

$+ \binom{12}{3} \times 1^9 \times \left(\frac{x}{4}\right)^3 + \ldots$

$= 1 + 3x + \frac{33}{8}x^2 + \frac{55}{16}x^3 + \ldots$

(b) $x = 0.1$ so $\left(1 + \frac{x}{4}\right)^{12} = (1.025)^{12}$

$(1.025)^{12} \approx 1 + 3 \times (0.1) + \frac{33}{8} \times (0.1)^2 + \frac{55}{16} \times (0.1)^3 + \ldots$

$= 1.3447$ (4 d.p.)

26. Proof

1　(a) No statement of proof at the end.

(b) $f(-2) = 0$ so, by the factor theorem, $(x + 2)$ is a factor of $f(x)$.

2　(a) $2(1)^2 + 11 = 13$; prime

$2(2)^2 + 11 = 19$; prime

$2(3)^2 + 11 = 29$; prime

$2(4)^2 + 11 = 43$; prime

$2(5)^2 + 11 = 61$; prime

$2(6)^2 + 11 = 83$; prime

$2(7)^2 + 11 = 109$; prime

$2(8)^2 + 11 = 139$; prime

$2(9)^2 + 11 = 173$; prime

$2(10)^2 + 11 = 211$; prime

(b) e.g. $n = 17$: $n^2 + n + 17 = (17)^2 + 17 + 17 = 323 = 17 \times 19$

3 Equivalently, need to prove that $(4-x)^2 - 7 + 2x \geqslant 0$ for all real values of x.

$$(4-x)^2 - 7 + 2x = 16 - 8x + x^2 - 7 + 2x$$
$$= x^2 - 6x + 9$$
$$= (x-3)^2$$

$(x-3)^2 \geqslant 0$ for all real values of x

So $(4-x)^2 - 7 + 2x \geqslant 0$ for all real values of x

So $(4-x)^2 \geqslant 7 - 2x$ for all real values of x, as required.

27. Cosine rule

(a) $a^2 = b^2 + c^2 - 2bc \cos A$

$PR^2 = 26^2 + 15^2 - 2 \times 26 \times 15 \times \cos 50°$
$\quad\ = 399.6256...$
$PR = 20.0\,\text{cm}$ (3 s.f.)

(b) $\cos \angle PQR = \dfrac{QR^2 + QP^2 - PR^2}{2 \times QR \times QP}$

$\qquad\qquad = \dfrac{20^2 + 13^2 - 20.0^2}{2 \times 20 \times 13} = 0.3257\ldots$

$\angle PQR = \cos^{-1}(0.3257\ldots) = 71.0°$ (1 d.p.)

28. Sine rule

(a) $\angle CAB = 180 - 60 - 36 = 84°$

$\dfrac{AC}{\sin B} = \dfrac{BC}{\sin A}$

$\dfrac{AC}{\sin 60°} = \dfrac{13}{\sin 84°}$

$\quad AC = \dfrac{13 \sin 60°}{\sin 84°} = 11.3$ cm (3 s.f.)

(b) Area $= \dfrac{1}{2} ab \sin C$

$\qquad\ = \dfrac{1}{2} \times 11.3 \times 13 \times \sin 36°$
$\qquad\ = 43.3\,\text{cm}^2$ (3 s.f.)

29. Trigonometric graphs

(i) (a)

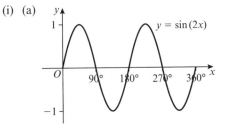

(b) $(0, 0)$, $(90°, 0)$, $(180°, 0)$, $(270°, 0)$, $(360°, 0)$

(ii) (a)

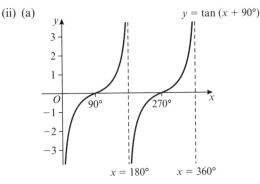

(b) $(90°, 0)$, $(270°, 0)$; asymptotes at $x = 0$, $x = 180°$ and $x = 360°$

30. Trigonometric equations 1

1 (a)

(b) $\sin^{-1}(-0.3) = -17.5°$ (1 d.p.)

$-17.5 + 360 = 342.5$

$180 - (-17.5) = 197.5$

$\qquad x = 342.5°, 197.5°$ (1 d.p.)

2 (a) $3 \cos \theta = 1$

$\cos \theta = \dfrac{1}{3}$

$\cos^{-1}\left(\dfrac{1}{3}\right) = 70.5°$ (1 d.p)

$\theta = 70.5°, -70.5°$ (1 d.p.)

(b) $\tan \theta + 2 = 0$

$\tan \theta = -2$

$\tan^{-1}(-2) = -63.4°$ (1 d.p.)

$-63.4, + 180 = 116.6$

$\theta = -63.4, 116.6°$ (1 d.p.)

31. Trigonometric identities

1 $3 \cos^2 x - 9 = 11 \sin x$

$3(1 - \sin^2 x) - 9 = 11 \sin x$

$3 - 3 \sin^2 x - 9 = 11 \sin x$

$3 \sin^2 x + 11 \sin x + 6 = 0$

$(3 \sin x + 2)(\sin x + 3) = 0$

$\sin x$ cannot equal -3, so $\sin x = -\dfrac{2}{3}$

$\sin^{-1}\left(-\dfrac{2}{3}\right) = -41.810\ldots°$

$-41.810\ldots + 360 = 318.189\ldots$

$180 - (-41.810\ldots) = 221.810\ldots$

$\qquad x = 221.8°, 318.9°$ (1 d.p.)

2 (a) $5 \sin x = 2 \tan x$

$5 \sin x = \dfrac{2 \sin x}{\cos x}$

$5 \sin x \cos x = 2 \sin x$

$5 \sin x \cos x - 2 \sin x = 0$

$\sin x (5 \cos x - 2) = 0$

(b) $\sin x = 0 \Rightarrow x = 0°, 180°$

$5 \cos x - 2 = 0 \Rightarrow \cos x = \dfrac{2}{5}$

$\cos^{-1}\left(\dfrac{2}{5}\right) = 66.42\ldots°$

$360 - 66.42\ldots = 293.57\ldots$

$x = 0°, 180°, 66.4°, 293.6°$ (1 d.p.)

32. Trigonometric equations 2

1 (a) $\sin(x - 40°) = -\dfrac{1}{2}$, $0 \leqslant x < 360°$

$-40° \leqslant x - 40° < 320°$

$\sin^{-1}\left(-\dfrac{1}{2}\right) = -30°$

$180° - (-30°) = 210°$

$x - 40° = -30°, 210°$

$x = 10°, 250°$

(b) $\cos(2x) = \frac{\sqrt{3}}{2}$, $0 \leqslant x < 360°$

$0 \leqslant 2x < 720°$

$\cos^{-1}\left(\frac{\sqrt{3}}{2}\right) = 30°$

$30 + 360 = 390$

$360 - 30 = 330$

$330 + 360 = 690$

$2x = 30°, 330°, 390°, 690°$

$x = 15°, 165°, 195°, 345°$

2 $\cos^2(x + 30°) = \frac{1}{4}$

$\cos(x + 30°) = \frac{1}{2}$ or $\cos(x + 30°) = -\frac{1}{2}$

$x + 30° = 60°, -60°$ $x + 30° = 120°, -120°$

$x \quad = 30°, -90°$ $x \quad = 90°, -150°$

So $x = -150°, -90°, 30°, 90°$.

33. Vectors

1 (a) $\overrightarrow{AB} = (5\mathbf{i} + 2\mathbf{j}) - (4\mathbf{i} - 2\mathbf{j}) = \mathbf{i} + 4\mathbf{j}$

   (b) $\overrightarrow{BA} = -\mathbf{i} - 4\mathbf{j}$

2 $\left|\binom{2}{-10}\right| = \sqrt{2^2 + 10^2} = 2\sqrt{26}$

$\frac{1}{2\sqrt{26}}\binom{2}{-10} = \begin{pmatrix} \frac{1}{\sqrt{26}} \\ -\frac{5}{\sqrt{26}} \end{pmatrix}$

34. Solving vector problems

1 (a) $\overrightarrow{AB} = \binom{8}{3} - \binom{0}{4} = \binom{8}{-1}$

$\overrightarrow{BC} = \binom{-2}{5} - \binom{8}{3} = \binom{-10}{2}$

$\overrightarrow{CA} = \binom{0}{4} - \binom{-2}{5} = \binom{2}{-1}$

(b) $|\overrightarrow{AB}| = \sqrt{8^2 + 1^2} = \sqrt{65}$

$|\overrightarrow{BC}| = \sqrt{10^2 + 2^2} = \sqrt{104}$

$|\overrightarrow{CA}| = \sqrt{2^2 + 1^2} = \sqrt{5}$

$\cos\angle ABC = \frac{|\overrightarrow{AB}|^2 + |\overrightarrow{BC}|^2 - |\overrightarrow{CA}|^2}{2|\overrightarrow{AB}||\overrightarrow{BC}|}$

$= \frac{65 + 104 - 5}{2\sqrt{65}\sqrt{104}}$

$= 0.997\ldots$

$\angle ABC = 4.184\ldots°$

Area of triangle $= \frac{1}{2}|\overrightarrow{AB}||\overrightarrow{BC}|\sin\angle ABC = 3$ units^2

(c) 6 units^2

(d) Option 1: $|\overrightarrow{OD}| = |\overrightarrow{OA}| + |\overrightarrow{BC}| = \binom{0}{4} + \binom{-10}{2} = \binom{-10}{6}$

Option 2: $|\overrightarrow{OD}| = |\overrightarrow{OB}| + |\overrightarrow{CA}| = \binom{8}{3} + \binom{2}{-1} = \binom{10}{2}$

Option 3: $|\overrightarrow{OB}| = |\overrightarrow{OC}| + |\overrightarrow{AB}| = \binom{-2}{5} + \binom{8}{-1} = \binom{6}{4}$

2 (a) $\overrightarrow{AB} = -2\mathbf{a} + 4\mathbf{b}$

$\overrightarrow{HA} = -\mathbf{b} + 2\mathbf{a}$ or $2\mathbf{a} - \mathbf{b}$

$\overrightarrow{JS} = 2\mathbf{b} - \frac{2}{3}(-2\mathbf{a} + 4\mathbf{b})$

$= 2\mathbf{b} + \frac{4\mathbf{a}}{3} - \frac{8\mathbf{b}}{3} = \frac{4\mathbf{a}}{3} - \frac{2\mathbf{b}}{3} = \frac{2}{3}(2\mathbf{a} - \mathbf{b})$

$\overrightarrow{KT} = \mathbf{b} - \frac{1}{3}(-2\mathbf{a} + 4\mathbf{b})$

$= \mathbf{b} + \frac{2\mathbf{a}}{3} - \frac{4\mathbf{b}}{3} = \frac{2\mathbf{a}}{3} - \frac{1\mathbf{b}}{3} = \frac{1}{3}(2\mathbf{a} - \mathbf{b})$

$\overrightarrow{HA}$, $\overrightarrow{JS}$ and $\overrightarrow{KT}$ are all multiples of $(2\mathbf{a} - \mathbf{b})$, so are parallel.

(b) $KT : JS : HA = 1 : 2 : 3$

35. Differentiating from first principles

1 $f(x) = x^2$

$f'(7) = \lim_{h \to 0} \frac{f(7 + h) - f(7)}{h}$

$= \lim_{h \to 0} \frac{(7 + h)^2 - 7^2}{h}$

$= \lim_{h \to 0} \frac{49 + 14h + h^2 - 49}{h}$

$= \lim_{h \to 0} (14 + h)$

As $h \to 0$, this limit tends to 14, so $f'(7) = 14$

2 (a) Gradient of $AB = \frac{(2(5 + h)^2 + 3(5 + h)) - (2(5)^2 + 3(5))}{h}$

$= \frac{50 + 20h + 2h^2 + 15 + 3h - 50 - 15}{h}$

$= \frac{23h + 2h^2}{h}$

$= 23 + 2h$ as required

(b) As $h \to 0$, then $2h \to 0$.

So gradient of tangent to curve at $A = 23$.

36. Differentiation 1

1 $y = \frac{x^2 + 6x + 9}{x} = x + 6 + 9x^{-1}$

$\frac{dy}{dx} = 1 - 9x^{-2}$

2 (a) $\frac{2 + 5\sqrt{x}}{x} = 2x^{-1} + 5x^{-\frac{1}{2}}$

(b) $y = 3x^2 + 1 - 2x^{-1} - 5x^{-\frac{1}{2}}$

$\frac{dy}{dx} = 6x + 2x^{-2} + \frac{5}{2}x^{-\frac{3}{2}}$

37. Differentiation 2

1 (a) $f'(x) = 9x^2 + 5$

(b) $9x^2 + 5 = 41$

$9x^2 = 36$

$x^2 = 4$

$x = \pm 2$

So $x = 2$ because $x > 0$.

2 $\frac{dy}{dx} = 4x - 8x^{-3}$

$\frac{d^2y}{dx^2} = 4 + 24x^{-4}$

3 (a) $y = x^3 + 2x^2 - 3x$

$\frac{dy}{dx} = 3x^2 + 4x - 3$

(b)

$y = x(x - 1)(x + 3)$

(c) At $(-3, 0)$: $\frac{dy}{dx} = 3(-3)^2 + 4(-3) - 3 = 27 - 12 - 3 = 12$

At $(0, 0)$: $\frac{dy}{dx} = 3(0)^2 + 4(0) - 3 = -3$

At $(1, 0)$: $\frac{dy}{dx} = 3(1)^2 + 4(1) - 3 = 4$

38. Tangents and normals

$\dfrac{dy}{dx} = \sqrt{4} + \dfrac{8}{4^2} - 5 = 2 + \dfrac{1}{2} - 5 = -\dfrac{5}{2}$

Gradient of tangent $= -\dfrac{5}{2}$

Gradient of normal $= \dfrac{2}{5}$

$\qquad y - y_1 = m(x - x_1)$

$\qquad y - 11 = \dfrac{2}{5}(x - 4)$

$\qquad 5y - 55 = 2x - 8$

$2x - 5y + 47 = 0$

39. Stationary points 1

1 $y = x^2 - 8x + 3$

$\dfrac{dy}{dx} = 2x - 8$

When $\dfrac{dy}{dx} = 0$, $2x - 8 = 0$

$\qquad\qquad\qquad x = 4$

$y = 4^2 - 8(4) + 3$

$\quad = 16 - 32 + 3 = -13$

Turning point is at $(4, -13)$.

2 $y = x^3 - 5x^2 + 8x + 1$

$\dfrac{dy}{dx} = 3x^2 - 10x + 8$

When $\dfrac{dy}{dx} = 0$, $3x^2 - 10x + 8 = 0$

$\qquad\qquad (3x - 4)(x - 2) = 0$

Stationary points at $x = \dfrac{4}{3}$ and $x = 2$.

40. Stationary points 2

(a) $y = 5x^2 - 3x - x^3$

$\dfrac{dy}{dx} = 10x - 3 - 3x^2$

When $\dfrac{dy}{dx} = 0$, $10x - 3 - 3x^2 = 0$

$\qquad\qquad 3x^2 - 10x + 3 = 0$

$\qquad\qquad (3x - 1)(x - 3) = 0$

A: $x = \dfrac{1}{3}$, $y = 5(\dfrac{1}{3})^2 - 3(\dfrac{1}{3}) - (\dfrac{1}{3})^3 = -\dfrac{13}{27}$, coordinates $(\dfrac{1}{3}, -\dfrac{13}{27})$

B: $x = 3$, $y = 5(3)^2 - 3(3) - 3^3 = 9$, coordinates $(3, 9)$

(b) $\dfrac{d^2y}{dx^2} = 10 - 6x$

At B, $x = 3$, so $\dfrac{d^2y}{dx^2} = 10 - 6 \times 3 = -8$

$\dfrac{d^2y}{dx^2} < 0$ so B is a maximum.

41. Modelling with calculus

1 (a) $P = 80x - \dfrac{x^2}{50}$

$\dfrac{dP}{dx} = 80 - \dfrac{x}{25}$

(b) When $\dfrac{dP}{dx} = 0$, $80 - \dfrac{x}{25} = 0$

$\qquad\qquad\qquad 80 = \dfrac{x}{25}$

$\qquad\qquad\qquad x = 2000$

$\dfrac{d^2P}{dx^2} = -\dfrac{1}{25} < 0$

Hence P is a maximum at $x = 2000$.

2 (a) $\pi r^2 x = 100$

$\qquad x = \dfrac{100}{\pi r^2}$

$\qquad A = \pi r^2 + 2\pi rx$

$\qquad\quad = \pi r^2 + 2\pi r\left(\dfrac{100}{\pi r^2}\right)$

$\qquad\quad = \pi r^2 + \dfrac{200}{r}$

(b) $A = \pi r^2 + 200r^{-1}$

$\dfrac{dA}{dr} = 2\pi r - \dfrac{200}{r^2}$

When $\dfrac{dA}{dr} = 0$, $2\pi r - \dfrac{200}{r^2} = 0$

$\qquad\qquad\qquad 2\pi r = \dfrac{200}{r^2}$

$\qquad\qquad\qquad 2\pi r^3 = 200$

$\qquad\qquad\qquad r^3 = \dfrac{100}{\pi}$

$\qquad\qquad\qquad r = \sqrt[3]{\dfrac{100}{\pi}} = 3.17$ (3 s.f.)

(c) $\dfrac{d^2A}{dr^2} = 2\pi + \dfrac{400}{r^3}$

When $r = 3.17$, $\dfrac{d^2A}{dr^2} = 2\pi + \dfrac{400}{3.17^3} = 18.84... > 0$

So A is a minimum.

(d) $A = \pi r^2 + \dfrac{200}{r} = \pi(3.17)^2 + \dfrac{200}{3.17} = 94.7\,\text{m}^2$ (3 s.f.)

42. Integration

1 $x - \dfrac{3x^4}{4} + c = x - \dfrac{3}{4}x^4 + c$

2 $\displaystyle\int(9x^2 + 6x + 1)\,dx = \dfrac{9x^3}{3} + \dfrac{6x^2}{2} + x + c$

$\qquad\qquad\qquad\qquad = 3x^3 + 3x^2 + x + c$

3 $y = 6x^2 + 5x^{\frac{3}{2}}$

$\displaystyle\int(6x^2 + 5x^{\frac{3}{2}})\,dx = \dfrac{6x^3}{3} + \dfrac{5x^{\frac{5}{2}}}{\left(\frac{5}{2}\right)} + c$

$\qquad\qquad\qquad\quad = 2x^3 + 2x^{\frac{5}{2}} + c$

43. Finding the constant

1 $f'(x) = 3 - \dfrac{2}{x^2} - \dfrac{3\sqrt{x}}{x^2}$

$\qquad\quad = 3 - 2x^{-2} - 3x^{-\frac{3}{2}}$

$f(x) = 3x - \dfrac{2x^{-1}}{-1} - \dfrac{3x^{-\frac{1}{2}}}{\left(-\frac{1}{2}\right)} + c$

$\qquad = 3x + \dfrac{2}{x} + \dfrac{6}{\sqrt{x}} + c$

$f(4) = 17$

So $3(4) + \dfrac{2}{4} + \dfrac{6}{\sqrt{4}} + c = 17$

$\qquad 12 + \dfrac{1}{2} + 3 + c = 17$

$\qquad\qquad\qquad c = \dfrac{3}{2}$

$f(x) = 3x + \dfrac{2}{x} + \dfrac{6}{\sqrt{x}} + \dfrac{3}{2}$

2 (a) $\dfrac{dy}{dx} = \dfrac{(x^2 + 5)^2}{x^2}$

$\qquad\quad = \dfrac{x^4 + 10x^2 + 25}{x^2}$

$\qquad\quad = x^2 + 10 + 25x^{-2}$

(b) $y = \dfrac{x^3}{3} + 10x + \dfrac{25x^{-1}}{-1} + c$

$\qquad = \dfrac{1}{3}x^3 + 10x - \dfrac{25}{x} + c$

At $x = 1$, $y = -13$:

$\dfrac{1}{3}(1)^3 + 10(1) - \dfrac{25}{(1)} + c = -13$

$\qquad \dfrac{1}{3} + 10 - 25 + c = -13$

$\qquad\qquad\qquad c = 2 - \dfrac{1}{3} = \dfrac{5}{3}$

$y = \dfrac{1}{3}x^3 + 10x - \dfrac{25}{x} + \dfrac{5}{3}$

44. Definite integration

1 $\displaystyle\int_1^3\left(3x^2 - 7 + \dfrac{6}{x^2}\right)dx = \left[x^3 - 7x - \dfrac{6}{x}\right]_1^3$

$\qquad\qquad\qquad = (27 - 21 - 2) - (1 - 7 - 6)$

$\qquad\qquad\qquad = 4 - (-12)$

$\qquad\qquad\qquad = 16$

2 $\int_1^2 \left(6x^{-3} - 2x^{-\frac{1}{2}}\right) dx = \left[\frac{-3}{x^2} - 4\sqrt{x}\right]_1^2$

$$= \left(\frac{-3}{4} - 4\sqrt{2}\right) - (-3 - 4)$$

$$= \frac{25}{4} - 4\sqrt{2}$$

45. Area under a curve

$y = x^3 - 6x^2 + 8x$

A_1 above the x-axis:

$\int_1^2 (x^3 - 6x^2 + 8x) dx = \left[\frac{1}{4}x^4 - 2x^3 + 4x^2\right]_1^2$

$$= (4 - 16 + 16) - \left(\frac{1}{4} - 2 + 4\right)$$

$$= 1.75$$

A_2 below the x-axis:

$\int_2^4 (x^3 - 6x^2 + 8x) dx = \left[\frac{1}{4}x^4 - 2x^3 + 4x^2\right]_2^4$

$$= (64 - 128 + 64) - (4 - 16 + 16)$$

$$= -4$$

So $A_2 = 4$

Total area $= 1.75 + 4 = 5.75$

46. More areas

(a) $5x - x^2 = x$

$5 - x = 1, x \neq 0$

$x = 4$

Coordinates of A are $(4, 4)$.

(b) Area of R = area under curve − area of triangle

Area under curve:

$\int_0^4 (5x - x^2) dx = \left[\frac{5}{2}x^2 - \frac{1}{3}x^3\right]_0^4$

$$= \left(40 - \frac{64}{3}\right) - (0)$$

$$= 18\frac{2}{3}$$

Area of triangle $= \frac{1}{2} \times 4 \times 4 = 8$

Area of $R = 18\frac{2}{3} - 8 = 10\frac{2}{3}$

47. Exponential functions

1 (a) C   (b) B   (c) D   (d) A

2 (a) $-2e^{-2x}$ (b) $10e^{5x}$ (c) $\frac{1}{3}e^{\frac{1}{3}x}$

48. Logarithms

1 (a) $y = 3^{-1} = \frac{1}{3}$

(b) $p^3 = 8 \Rightarrow p = 2$

(c) $\log_4 8 = \frac{3}{2}$

2 (a) $\log_a (5^2) = \log_a 25$

(b) $\log_a (2 \times 9) = \log_a 18$

(c) $\log_a (4^3) - \log_a 8 = \log_a \left(\frac{4^3}{8}\right) = \log_a 8$

3 $2\log_8 6 - \log_8 9 = \log_8 (6^2) - \log_8 9$

$$= \log_8 \left(\frac{6^2}{9}\right) = \log_8 4$$

$$= \frac{2}{3}$$

since $8^{\frac{2}{3}} = \left(\sqrt[3]{8}\right)^2 = 4$

49. Equations with logs

1 $\log_2 (x + 1) - \log_2 x = \log_2 5$

$$\log_2 \left(\frac{x + 1}{x}\right) = \log_2 5$$

$$\frac{x + 1}{x} = 5$$

$$x + 1 = 5x$$

$$x = \frac{1}{4}$$

2 $\log_6 (x - 1) + \log_6 x = 1$

$\log_6 [(x - 1)x] = 1$

$x(x - 1) = 6$

$x^2 - x - 6 = 0$

$(x + 2)(x - 3) = 0$

$x = 3$ because $\log_6 (-2)$ is not defined.

3 $\log_3 (x - 1) = -1$

$x - 1 = \frac{1}{3}$

$x = \frac{4}{3}$

4 $2\log_4 x - \log_4 (x - 3) = 2$

$\log_4 (x^2) - \log_4 (x - 3) = 2$

$\log_4 \left(\frac{x^2}{x - 3}\right) = 2$

$\frac{x^2}{x - 3} = 16$

$x^2 = 16x - 48$

$x^2 - 16x + 48 = 0$

$(x - 12)(x - 4) = 0$

$x = 12$ or $x = 4$

50. Exponential equations

1 (a) $2^b = 15$

$b = \log_2 15 = 3.91$ (3 s.f.)

(b) $6^x = 0.4$

$x = \log_6 0.4 = -0.511$ (3 s.f.)

2 (a) $\quad 3^{2x} + 3^x = 6$

$(3^x)^2 + 3^x - 6 = 0$

$(3^x - 2)(3^x + 3) = 0$

$3^x = 2 \Rightarrow x = \log_3 2 = 0.63$ (2 d.p.)

(b) The other factor gives $3^x = -3$, but 3^x is positive for all values of x so this solution does not exist.

3 $\quad\quad 3^{x-1} = 2^{2x+1}$

$(x - 1)\log 3 = (2x + 1) \log 2$

$x \log 3 - 2x \log 2 = \log 2 + \log 3$

$x (\log 3 - 2 \log 2) = \log 2 + \log 3$

$$x = \frac{\log 2 + \log 3}{\log 3 - 2 \log 2}$$

$$= -6.23 \text{ (3 s.f.)}$$

4 If curves intersect then x-coordinate satisfies:

$$6^{x^2} = 2^{x-1}$$

$$x^2 \log 6 = (x - 1) \log 2$$

$(\log 6) x^2 - (\log 2) x + \log 2 = 0$

Discriminant $= (\log 2)^2 - 4(\log 6)(\log 2)$

$$= -0.8463\ldots$$

So no real solutions, so curves do not intersect.

51. Natural logarithms

1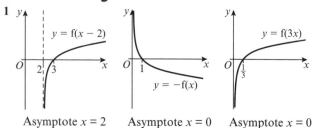

Asymptote $x = 2$ Asymptote $x = 0$ Asymptote $x = 0$

2 $6 = 3\,e^{2x-1}$
$2 = e^{2x-1}$
$\ln 2 = 2x - 1$
$x = \frac{1}{2} + \frac{1}{2}\ln 2$

3 (a) $\ln\left(\dfrac{x+1}{x}\right) = \ln 5$

$\dfrac{x+1}{x} = 5$

$x + 1 = 5x$

$x = \frac{1}{4}$

(b) $(e^{2x})^2 + 3\,e^{2x} - 10 = 0$
$(e^{2x} + 5)(e^{2x} - 2) = 0$
$\cancel{e^{2x} = -5}$ or $e^{2x} = 2$
$2x = \ln 2$
$x = \frac{1}{2}\ln 2$

(c) $\ln(6x + 7) = \ln x^2$
$6x + 7 = x^2$
$x^2 - 6x - 7 = 0$
$(x - 7)(x + 1) = 0$
$x = 7$ or $\cancel{x = -1}$

4 $\ln(3^x\,e^{2x-5}) = \ln 7$
$\ln 3^x + \ln(e^{2x-5}) = \ln 7$
$x \ln 3 + 2x - 5 = \ln 7$
$x(2 + \ln 3) = 5 + \ln 7$
$x = \dfrac{5 + \ln 7}{2 + \ln 3}$

5 (a) $f(x) = \dfrac{3x^2 - 7x + 2}{x^2 - 4}$

$= \dfrac{(3x - 1)(x - 2)}{(x + 2)(x - 2)}$

$= \dfrac{3x - 1}{x + 2}$

(b) $\ln(3x^2 - 7x + 2) - \ln(x^2 - 4) = 1$

$\ln\left(\dfrac{3x^2 - 7x + 2}{x^2 - 4}\right) = 1$

$\dfrac{3x^2 - 7x + 2}{x^2 - 4} = e$

$\dfrac{3x - 1}{x + 2} = e$

$3x - 1 = ex + 2e$
$3x - ex = 1 + 2e$
$x(3 - e) = 1 + 2e$
$x = \dfrac{1 + 2e}{3 - e}$

52. Exponential modelling

1 (a) 100

(b) $N = 100\,e^{0.8t}$

$\dfrac{dN}{dt} = 0.8 \times 100\,e^{0.8t}$

$= 0.8N$

2 (a) 250 grams

(b) $125 = 250\,e^{-90k}$
$e^{-90k} = 0.5$
$-90k = \ln 0.5$
$k = 0.00770$ (3 s.f.)

53. Modelling with logs

(a) $x = aN^b$

$\log x = \log(aN^b)$

$= \log(N^b) + \log a$

$= b \log N + \log a$

So $k = b$ and $c = \log a$

(b)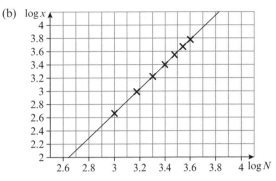

Graph of $\log x$ against $\log N$ is approximately linear so the model is accurate.

(c) Equation of graph is approximately
$\log x = 1.85 \log N - 2.9$, so $k = 1.85$ and $c = -2.9$
So $a = 10^{-2.9} = 0.0013$ (2 s.f.) and $b = 1.85$

STATISTICS

56. Sampling

(a) $152 \times \dfrac{50}{436} = 17.43...$
17 female employees from Bristol

(b) A numbered list of all the employees

57. Mean

(a) $251 + 19 + 22 + 15 + 21 + 16 = 344$

$\dfrac{344}{20} = 17.2\,°C$

(b) The mean will go down because $16 < 17.2$.

58. Median and quartiles

(a) $22 \div 2 = 11$, so median is half way between the 11th and 12th data values
Median = 1016.5 hPa

(b) $22 \div 4 = 5.5$ so Q_1 is the 6th data value
$Q_1 = 1012$ hPa
$(3 \times 22) \div 4 = 16.5$ so Q_3 is the 17th data value
$Q_3 = 1026$ hPa
IQR $= Q_3 - Q_1 = 1026 - 1012 = 14$ hPa

59. Linear interpolation

$\dfrac{178}{4} = 44.5$: $Q_1 = 14.5 + (44.5 - 29) \times \dfrac{3}{64} = 15.23$ (2 d.p.)

$\dfrac{178}{2} = 89$: $Q_2 = 14.5 + (89 - 29) \times \dfrac{3}{64} = 17.31$ (2 d.p.)

$0.9 \times 178 = 160.2$;
90th percentile $= 19.5 + (160.2 - 148) \times \dfrac{3}{21}$
$= 21.24$ (2 d.p.)

60. Standard deviation 1

(a) $n = 31$

$\bar{x} = \dfrac{397.3}{31} = 12.8\,°C$ (1 d.p.)

$\sigma = \sqrt{\dfrac{5165.39}{31} - \left(\dfrac{397.3}{31}\right)^2} = 1.5\,°C$ (1 d.p.)

(b) E.g. No, the mean temperature in Leuchars was higher and Leuchars is further north than Camborne *or* No, this is not a sufficiently large sample

61. Standard deviation 2

Midpoint	2	5	8.5	15.5
Frequency	48	31	15	6

Mean = 4.7 mins (1 d.p.)

Standard deviation = 3.6 mins (1 d.p.)

62. Coding

1 $\bar{y} = \dfrac{4380}{184} = 23.804\ldots$

So $\bar{x} = 100 \times 23.804\ldots = 2380\,\text{m}$ (3 s.f.)

$\sigma_y = \sqrt{\dfrac{2727.3}{184}} = 3.849\ldots$

So $\sigma_x = 100 \times 3.849\ldots = 385\,\text{m}$ (3 s.f.)

2 (a) $\bar{x} = \dfrac{668}{10} = 66.8$

$\sigma_x = \sqrt{\dfrac{47870}{10} - \left(\dfrac{668}{10}\right)^2} = 18.0$ (1 d.p.)

(b) $y = 1.1(x - 8)$

$\bar{y} = 1.1(66.8 - 8) = 64.7$ (1 d.p.)

$\sigma_y = 1.1 \times 18.0 = 19.8$ (1 d.p.)

63. Box plots and outliers

(a) (i) 20 m

(ii) Upper quartile (Q_3)

(b) It is an outlier. It is a non-typical data value that is usually more than $1.5 \times (Q_3 - Q_1)$ above Q_3.

64. Cumulative frequency diagrams

(a)

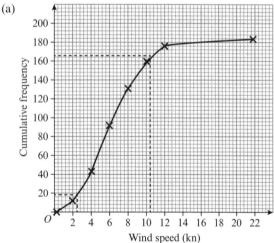
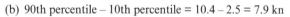

(b) 90th percentile – 10th percentile = 10.4 – 2.5 = 7.9 kn

65. Histograms

(a) 63.5 – 57.5 = 6

57.5 – 54.5 = 3

55–57 bar will be 2 cm wide.

(b) Area of $6 \times 4 = 24\,\text{cm}^2$ represents 30 apples.

$\dfrac{24}{30} \times 12 = 9.6$, so 12 apples represented by area of $9.6\,\text{cm}^2$.

$\dfrac{9.6}{2} = 4.8$ so bar is 4.8 cm high.

66. Comparing distributions

(a) 1987:

$\bar{x} = \dfrac{710.9}{31} = 22.9°\,\text{C}$ (1 d.p.)

$\sigma_x = \sqrt{\dfrac{16\,364.77}{31} - \left(\dfrac{710.9}{31}\right)^2} = 1.4\,°\text{C}$ (1 d.p.)

2015:

$\bar{x} = \dfrac{727.9}{31} = 3.5\,°\text{C}$ (1 d.p.)

$\sigma_x = \sqrt{\dfrac{17\,255.61}{31} - \left(\dfrac{729.9}{31}\right)^2} = 2.3°\text{C}$ (1 d.p.)

(b) E.g. Mean temperatures have increased slightly (increase in mean from 22.9°C to 23.5°C) but the temperatures recorded in May 2015 were also more varied (increase in s.d. from 1.4°C to 2.3°C). This is also a relatively small sample so it is not sufficient evidence to conclude that average air temperatures have increased between 1987 and 2015.

67. Correlation and cleaning data

(a) The group were all young people so it is likely that the age of 81 is an error. Michelle's decision is valid.

(b)

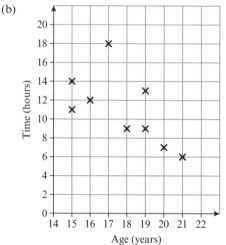

(c) Negative correlation. The older the person is, the fewer hours of training are likely to be needed to achieve the standard of proficiency.

68. Regression

(a) For every unit (hPa) increase in pressure, the visibility increases by approximately 0.373 km.

(b) E.g. The scatter diagram shows that there is only very weak linear correlation between visibility and mean pressure, so a linear regression model is unlikely to model the data accurately.

69. Using regression lines

(a) The data shows strong (positive) linear correlation.

(b) (i) This estimate is from within the range of the independent variable (interpolation) so is reliable.

(ii) This estimate is from outside the range of the independent variable (extrapolation) so is unreliable.

70. Drawing Venn diagrams

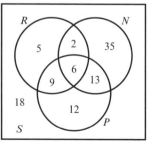

71. Using Venn diagrams

(a)
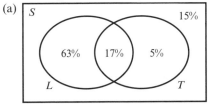

(b) P(L and T') = 0.63

72. Independent events

(a) A and C

(b) P(B and C) = P(B) × P(C)

$$x = (3x + 0.1 + x) \times (x + 0.1)$$
$$= (4x + 0.1)(x + 0.1)$$
$$= 4x^2 + 0.5x + 0.01$$
$$100x = 400x^2 + 50x + 1$$
$$0 = 400x^2 - 50x + 1$$

(c) (40x – 1)(10x – 1) = 0

x = 0.1 or x = 0.025

x = 0.1 would give

P(A or B or C) = 0.45 + 0.3 + 0.1 + 0.1 + 0.1
                = 1.05

which is greater than 1 and hence impossible.

So x = 0.025

y = 1 – (0.45 + 0.075 + 0.1 + 0.025 + 0.1)
  = 0.25

73. Tree diagrams

(a)
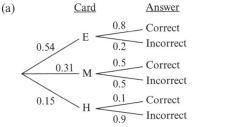

(b) P(Correct) = 0.54 × 0.8 + 0.31 × 0.5 + 0.15 × 0.1 = 0.602

74. Random variables

1 (a)

y	2	3	4	5
P(Y = y)	$\frac{1}{30}$	$\frac{2}{15}$	$\frac{3}{10}$	$\frac{8}{15}$

(b) P(Y > 3) = P(Y = 4) + P(Y = 5) = $\frac{3}{10} + \frac{8}{15} = \frac{5}{6}$

2 (a) 0.1 + a + 0.15 + 2a + 0.15 = 1

$$3a = 0.6$$
$$a = 0.2$$

(b) $P(3X + 1 \leq 6) = P(X \leq \frac{5}{3})$

$$= P(X = -2) + P(X = -1) + P(X = 0) + P(X = 1)$$
$$= 0.1 + 0.2 + 0.15 + 0.4$$
$$= 0.85$$

75. The binomial distribution

1 (a) There are a fixed number of rolls (50), and each roll is independent. The probabilities are the same on each roll.

(b) $X \sim B\left(50, \frac{1}{6}\right)$

(i) $P(X = 10) = \binom{50}{10}\left(\frac{1}{6}\right)^{10}\left(\frac{5}{6}\right)^{40} = 0.1156$ (4 d.p.)

(ii) Using calculator with p = 0.1667:

$$P(X < 7) = P(X \leq 6) = 0.2504 \text{ (4 d.p.)}$$

2 Because Emma is selecting without replacement, the probability of success is not the same for each of the trials.

76. Hypothesis testing

1 Assume H_0, so $X \sim B(28, 0.7)$

$$P(X \geq 25) = 1 - P(X \leq 24)$$
$$= 1 - 0.9843$$
$$= 0.0157$$

1.6% < 2% so there is enough evidence to reject H_0.

2 Let X = the number of students out of 45 who pass. Then $X \sim B(45, p)$.

H_0: p = 0.493, H_1: p ≠ 0.493

Assume H_0 is true, so $X \sim B(45, 0.493)$.

$$P(X \geq 29) = 1 - P(X \leq 28)$$
$$= 1 - 0.9706$$
$$= 0.0294$$

2.9% > 2.5% so there is not enough evidence to reject H_0.
Conclude that the average pass rate in Edinburgh was no different from the national average.

77. Critical regions

(a) Assume H_0 is true, so $X \sim B(15, 0.45)$

$$P(X \leq 2) = 0.0107$$
$$P(X \geq 12) = 1 - P(X \leq 11)$$
$$= 1 - 0.9937$$
$$= 0.0063$$

The critical regions are $X \leq 2$ and $X \geq 12$

(b) 0.0107 + 0.0063 = 0.017

Actual significance level = 0.017 or 1.7%

MECHANICS

80. Modelling in mechanics

(a) There is no frictional force between block P and the table top.

(b) Air resistance and rotational forces can be ignored.

(c) The magnitude of the acceleration is the same for both particles.

(d) The weight of the string can be ignored.

(e) The tension in the string is the same on both sides of the pulley.

81. Motion graphs

(a)
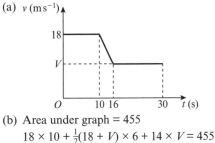

(b) Area under graph = 455

$$18 \times 10 + \frac{1}{2}(18 + V) \times 6 + 14 \times V = 455$$
$$180 + 54 + 3V + 14V = 455$$
$$17V = 221$$
$$V = 13$$

82. Constant acceleration 1

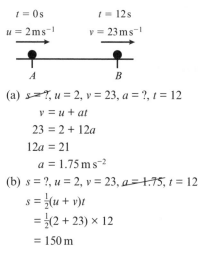

$t = 0\,\text{s}$ $t = 12\,\text{s}$
$u = 2\,\text{m s}^{-1}$ $v = 23\,\text{m s}^{-1}$

A B

(a) $s = ?, u = 2, v = 23, a = ?, t = 12$
$$v = u + at$$
$$23 = 2 + 12a$$
$$12a = 21$$
$$a = 1.75\,\text{m s}^{-2}$$

(b) $s = ?, u = 2, v = 23, a = 1.75, t = 12$
$$s = \tfrac{1}{2}(u + v)t$$
$$= \tfrac{1}{2}(2 + 23) \times 12$$
$$= 150\,\text{m}$$

83. Constant acceleration 2

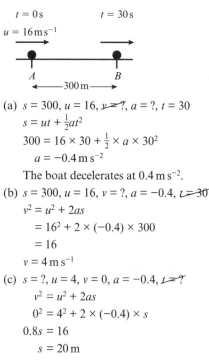

$t = 0\,\text{s}$ $t = 30\,\text{s}$
$u = 16\,\text{m s}^{-1}$

A ←300 m→ B

(a) $s = 300, u = 16, v = ?, a = ?, t = 30$
$$s = ut + \tfrac{1}{2}at^2$$
$$300 = 16 \times 30 + \tfrac{1}{2} \times a \times 30^2$$
$$a = -0.4\,\text{m s}^{-2}$$
The boat decelerates at $0.4\,\text{m s}^{-2}$.

(b) $s = 300, u = 16, v = ?, a = -0.4, t = 30$
$$v^2 = u^2 + 2as$$
$$= 16^2 + 2 \times (-0.4) \times 300$$
$$= 16$$
$$v = 4\,\text{m s}^{-1}$$

(c) $s = ?, u = 4, v = 0, a = -0.4, t = ?$
$$v^2 = u^2 + 2as$$
$$0^2 = 4^2 + 2 \times (-0.4) \times s$$
$$0.8s = 16$$
$$s = 20\,\text{m}$$

84. Motion under gravity

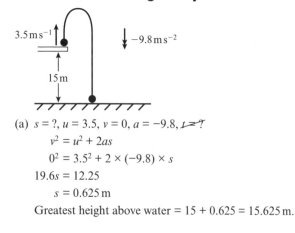

$3.5\,\text{m s}^{-1}$ $\downarrow -9.8\,\text{m s}^{-2}$

$15\,\text{m}$

(a) $s = ?, u = 3.5, v = 0, a = -9.8, t = ?$
$$v^2 = u^2 + 2as$$
$$0^2 = 3.5^2 + 2 \times (-9.8) \times s$$
$$19.6s = 12.25$$
$$s = 0.625\,\text{m}$$
Greatest height above water $= 15 + 0.625 = 15.625\,\text{m}$.

(b) $s = -15, u = 3.5, v = ?, a = -9.8, t = ?$
$$v^2 = u^2 + 2as$$
$$= 3.5^2 + 2 \times (-9.8) \times (-15)$$
$$= 306.25$$
$$v = \pm 17.5$$
The diver hits the water with speed $17.5\,\text{m s}^{-1}$.

(c) $s = -15, u = 3.5, v = -17.5, a = -9.8, t = ?$
$$v = u + at$$
$$-17.5 = 3.5 + (-9.8) \times t$$
$$9.8t = 21$$
$$t = 2.1428\ldots = 2.1\,\text{s (2 s.f.)}$$

85. Forces

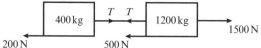

$200\,\text{N}$ 400 kg T T 1200 kg $\rightarrow$ 1500 N
$500\,\text{N}$

(a) Using $F = ma$ (whole system):
$$1500 - 500 - 200 = (400 + 1200)a$$
$$800 = 1600a$$
$$a = 0.5\,\text{m s}^{-2}$$

(b) Using $F = ma$ (trailer):
$$T - 200 = 400 \times 0.5$$
$$T = 400\,\text{N}$$

86. Forces as vectors

(a) $\mathbf{R} = \mathbf{F}_1 + \mathbf{F}_2 = (2\mathbf{i} - 6\mathbf{j}) + (3\mathbf{i} + k\mathbf{j})$
$$= [5\mathbf{i} + (-6 + k)\mathbf{j}]$$
$$5 = -(-6 + k)$$
$$k = 1$$

(b) $\mathbf{F}_1 + \mathbf{F}_2 + \mathbf{F}_3 = 0$
$(2\mathbf{i} - 6\mathbf{j}) + (3\mathbf{i} + \mathbf{j}) + (p\mathbf{i} + q\mathbf{j}) = 0$
$\mathbf{i}$ components: $2 + 3 + p = 0$, so $p = -5$
$\mathbf{j}$ components: $-6 + 1 + q = 0$, so $q = 5$

87. Motion in 2D

1 (a) $\mathbf{F} = m\mathbf{a}$
$$\begin{pmatrix} 24 \\ -10 \end{pmatrix} = 8\mathbf{a}$$
$$\mathbf{a} = \tfrac{1}{8}\begin{pmatrix} 24 \\ -10 \end{pmatrix} = \begin{pmatrix} 3 \\ -1.25 \end{pmatrix}\,\text{m s}^{-2}$$

(b) Magnitude $= \sqrt{3^2 + 1.25^2} = 3.25\,\text{m s}^{-2}$

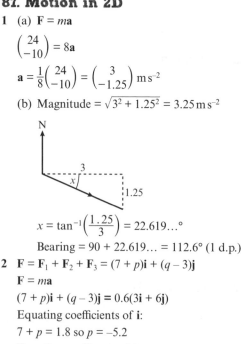

$$x = \tan^{-1}\left(\frac{1.25}{3}\right) = 22.619\ldots^\circ$$
Bearing $= 90 + 22.619\ldots = 112.6^\circ$ (1 d.p.)

2 $\mathbf{F} = \mathbf{F}_1 + \mathbf{F}_2 + \mathbf{F}_3 = (7 + p)\mathbf{i} + (q - 3)\mathbf{j}$
$\mathbf{F} = m\mathbf{a}$
$(7 + p)\mathbf{i} + (q - 3)\mathbf{j} = 0.6(3\mathbf{i} + 6\mathbf{j})$
Equating coefficients of $\mathbf{i}$:
$7 + p = 1.8$ so $p = -5.2$
Equating coefficents of $\mathbf{j}$:
$q - 3 = 3.6$ so $q = 6.6$

88. Pulleys

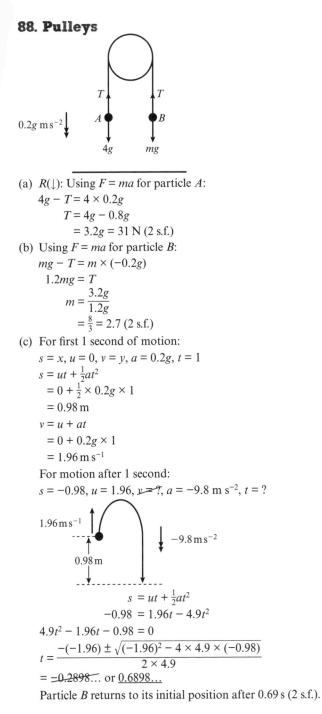

(a) $R(\downarrow)$: Using $F = ma$ for particle A:
$$4g - T = 4 \times 0.2g$$
$$T = 4g - 0.8g$$
$$= 3.2g = 31\text{ N (2 s.f.)}$$

(b) Using $F = ma$ for particle B:
$$mg - T = m \times (-0.2g)$$
$$1.2mg = T$$
$$m = \frac{3.2g}{1.2g}$$
$$= \tfrac{8}{3} = 2.7\text{ (2 s.f.)}$$

(c) For first 1 second of motion:
$$s = x, u = 0, v = y, a = 0.2g, t = 1$$
$$s = ut + \tfrac{1}{2}at^2$$
$$= 0 + \tfrac{1}{2} \times 0.2g \times 1$$
$$= 0.98\text{ m}$$
$$v = u + at$$
$$= 0 + 0.2g \times 1$$
$$= 1.96\text{ m s}^{-1}$$

For motion after 1 second:
$$s = -0.98, u = 1.96, v = ?, a = -9.8\text{ m s}^{-2}, t = ?$$

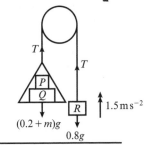

$$s = ut + \tfrac{1}{2}at^2$$
$$-0.98 = 1.96t - 4.9t^2$$
$$4.9t^2 - 1.96t - 0.98 = 0$$
$$t = \frac{-(1.96) \pm \sqrt{(-1.96)^2 - 4 \times 4.9 \times (-0.98)}}{2 \times 4.9}$$
$$= -0.2898\ldots \text{ or } 0.6898\ldots$$

Particle B returns to its initial position after 0.69 s (2 s.f.).

89. Connected particles

(a) $R(\uparrow)$: Using $F = ma$ for particle R:
$$T - 0.8g = 0.8 \times 1.5$$
$$T = 0.8g + 1.2$$
$$= 9.04\text{ N}$$

$R(\downarrow)$: Using $F = ma$ for particles P and Q combined:
$$(0.2 + m)g - T = (0.2 + m) \times 1.5$$
$$0.2g + mg - 9.04 = 0.3 + 1.5m$$
$$1.96 + 9.8m - 9.04 = 0.3 + 1.5m$$
$$8.3m = 7.38$$
$$m = 0.88915\ldots = 0.89\text{ (2 s.f.)}$$

(b) $R(\downarrow)$: Using $F = ma$ on block P:
$$0.2g - R = 0.2 \times 1.5$$
$$R = 0.2g - 0.3$$
$$= 1.66\text{ N}$$

So the force exerted on P by Q
is 1.7 N (2 s.f.).

90. Combining techniques

Consider system before tow rope breaks:
$$\text{Resultant force} = 4000 - (1000 + 1800) = 1200\text{ N}$$
$$1200 = (m_1 + m_2) \times 0.4$$
$$m_1 + m_2 = 3000$$

Consider trailer after tow rope breaks:
$$v = u + at$$
$$0 = 14 + 2.8a$$
$$a = -5\text{ ms}^{-1}$$
$$-1000 = m_1 \times (-5)$$
$$m_1 = 200$$
So $m_2 = 2800$

The trailer has a mass of 200 kg and the truck has a mass of 2800 kg

91. Variable acceleration 1

(a) $x = t^4 - 20t^3 + 96t^2$
$$= t^2(t^2 - 20t + 96)$$
$$= t^2(t - 12)(t - 8)$$

Both linear factors are $\geqslant 0$ for all values in the range $0 \leqslant t \leqslant 8$, and $t^2 \geqslant 0$ so $x \geqslant 0$ in this range.

(b) $\dfrac{dx}{dt} = 4t^3 - 60t^2 + 96t$
$$= 4t(t^2 - 15t + 24)$$
$\dfrac{dx}{dt} = 0 \Rightarrow 4t(t^2 - 15t + 24) = 0$, so either $t = 0$ or
$t^2 - 15t + 24 = 0 \Rightarrow t = 13.17\ldots$ or $t = 1.82\ldots$ (3 s.f.)

Maximum occurs at $t = 1.82\ldots$

At this point, $x = 209$ m (3 s.f.)

Graph is a positive quartic graph and $t = 1.82$ is the middle of three stationary points, so it must be a maximum.

92. Variable acceleration 2

(a) $a = \dfrac{dv}{dt} = \tfrac{1}{2} \times 0.1t^{-\frac{1}{2}} - 0.2$
$$= 0.05t^{-\frac{1}{2}} - 0.2$$
When $t = 4$, $a = 0.05(4)^{-\frac{1}{2}} - 0.2 = 0.025 - 0.2 = -0.175\text{ ms}^{-2}$

(b) $\dfrac{dx}{dt} = 0.1t^{\frac{1}{2}} - 0.2t$
$$x = \tfrac{2}{3} \times 0.1t^{\frac{3}{2}} - \tfrac{1}{2} \times 0.2t^2 + c$$
$$= \tfrac{1}{15}t^{\frac{3}{2}} - \tfrac{1}{10}t^2 + c$$
When $t = 0$, $x = 4 \Rightarrow c = 4$

So $x = \tfrac{1}{15}t^{\frac{3}{2}} - \tfrac{1}{10}t^2 + 4$

(c) When $t = 5$, $x = \tfrac{1}{15}(5)^{\frac{3}{2}} - \tfrac{1}{10}(5)^2 + 4 = 2.25$ m (3 s.f.)

93. Deriving the *suvat* equations

1 (a) $v = \dfrac{\mathrm{d}s}{\mathrm{d}t} = 10 - 2kt$

   $a = \dfrac{\mathrm{d}^2s}{\mathrm{d}t^2} = -2k$

   So a is constant.

   (b) $v = 0$ when $t = 4$ so

   $0 = 10 - 2 \times k \times 4$

   $8k = 10$

   $k = 1.25$

2 Distance travelled = area under graph

   $$= UT + \tfrac{1}{2}(V - U)T$$

   $$= T(U + \tfrac{1}{2}V - \tfrac{1}{2}U)$$

   $$= \left(\dfrac{U + V}{2}\right)T$$

   as required.

Published by Pearson Education Limited, 80 Strand, London, WC2R 0RL.
www.pearsonschoolsandfecolleges.co.uk

Copies of official specifications for all Pearson qualifications may be found on the website: qualifications.pearson.com

Text and illustrations © Pearson Education Ltd 2017
Typeset and illustrated by Techset
Produced by ProjectOne
Cover illustration by Miriam Sturdee

The right of Harry Smith to be identified as author of this work has been asserted by him in accordance with the Copyright, Designs and Patents Act 1988.

First published 2017

20 19 18 17
10 9 8 7 6 5 4 3 2 1

British Library Cataloguing in Publication Data
A catalogue record for this book is available from the British Library

ISBN 978 1 292 19066 2

Printed in Slovakia by Neografia

Notes from the publisher
1. While the publishers have made every attempt to ensure that advice on the qualification and its assessment is accurate, the official specification and associated assessment guidance materials are the only authoritative source of information and should always be referred to for definitive guidance.

Pearson examiners have not contributed to any sections in this resource relevant to examination papers for which they have responsibility.

2. Pearson has robust editorial processes, including answer and fact checks, to ensure the accuracy of the content in this publication, and every effort is made to ensure this publication is free of errors. We are, however, only human, and occasionally errors do occur. Pearson is not liable for any misunderstandings that arise as a result of errors in this publication, but it is our priority to ensure that the content is accurate. If you spot an error, please do contact us at resourcescorrections@pearson.com so we can make sure it is corrected.